Flavoprotein Protocols

METHODS IN MOLECULAR BIOLOGY™

John M. Walker, Series Editor

141. **Plant Hormone Protocols**, edited by *Jeremy A. Roberts and Gregory A. Tucker, 2000*
140. **Chaperonin Protocols**, edited by *Christine Schneider, 2000*
139. **Extracellular Matrix Protocols**, edited by *Charles Streuli and Michael Grant, 2000*
138. **Chemokine Protocols**, edited by *Amanda E. I. Proudfoot, Timothy N. C. Wells, and Christine Power, 2000*
137. **Developmental Biology Protocols, Volume III**, edited by *Rocky S. Tuan and Cecilia W. Lo, 2000*
136. **Adrenergic Receptor Protocols**, edited by *Curtis A. Machida, 2000*
135. **Glycoproteins Methods and Protocols:** *The Mucins*, edited by *Anthony P. Corfield, 2000*
134. **T Cell Protocols:** *Development and Activation*, edited by *Kelly P. Kearse, 2000*
133. **Gene Targeting Protocols,** edited by *Eric B. Kmiec, 1999*
132. **Bioinformatics Methods and Protocols,** edited by *Stephen Misener and Stephen A. Krawetz, 1999*
131. **Flavoprotein Protocols,** edited by *S. K. Chapman and G. A. Reid, 1999*
130. **Transcription Factor Protocols,** edited by *Martin J. Tymms, 1999*
129. **Integrin Protocols,** edited by *Anthony Howlett, 1999*
128. **NMDA Protocols,** edited by *Min Li, 1999*
127. **Molecular Methods in Developmental Biology:** Xenopus *and Zebrafish*, edited by *Matthew Guille, 1999*
126. **Developmental Biology Protocols, Volume II,** edited by *Rocky S. Tuan and Cecilia W. Lo, 2000*
125. **Developmental Biology Protocols, Volume I,** edited by *Rocky S. Tuan and Cecilia W. Lo, 2000*
124. **Protein Kinase Protocols,** edited by *Alastair Reith, 1999*
123. ***In Situ* Hybridization Protocols (2nd ed.),** edited by *Ian A. Darby, 1999*
122. **Confocal Microscopy Methods and Protocols,** edited by *Stephen W. Paddock, 1999*
121. **Natural Killer Cell Protocols:** *Cellular and Molecular Methods*, edited by *Kerry S. Campbell and Marco Colonna, 1999*
120. **Eicosanoid Protocols,** edited by *Elias A. Lianos, 1999*
119. **Chromatin Protocols,** edited by *Peter B. Becker, 1999*
118. **RNA–Protein Interaction Protocols,** edited by *Susan R. Haynes, 1999*
117. **Electron Microscopy Methods and Protocols,** edited by *M. A. Nasser Hajibagheri, 1999*
116. **Protein Lipidation Protocols,** edited by *Michael H. Gelb, 1999*
115. **Immunocytochemical Methods and Protocols (2nd ed.),** edited by *Lorette C. Javois, 1999*
114. **Calcium Signaling Protocols,** edited by *David G. Lambert, 1999*
113. **DNA Repair Protocols:** *Eukaryotic Systems*, edited by *Daryl S. Henderson, 1999*
112. **2-D Proteome Analysis Protocols,** edited by *Andrew J. Link 1999*
111. **Plant Cell Culture Protocols,** edited by *Robert D. Hall, 1999*
110. **Lipoprotein Protocols,** edited by *Jose M. Ordovas, 1998*
109. **Lipase and Phospholipase Protocols,** edited by *Mark H. Doolittle and Karen Reue, 1999*
108. **Free Radical and Antioxidant Protocols,** edited by *Donald Armstrong, 1998*
107. **Cytochrome P450 Protocols,** edited by *Ian R. Phillips and Elizabeth A. Shephard, 1998*
106. **Receptor Binding Techniques,** edited by *Mary Keen, 1999*
105. **Phospholipid Signaling Protocols,** edited by *Ian M. Bird, 1998*
104. **Mycoplasma Protocols,** edited by *Roger J. Miles and Robin A. J. Nicholas, 1998*
103. ***Pichia* Protocols,** edited by *David R. Higgins and James M. Cregg, 1998*
102. **Bioluminescence Methods and Protocols,** edited by *Robert A. LaRossa, 1998*
101. **Mycobacteria Protocols,** edited by *Tanya Parish and Neil G. Stoker, 1998*
100. **Nitric Oxide Protocols,** edited by *Michael A. Titheradge, 1998*
99. **Human Cytokines and Cytokine Receptors,** edited by *Reno Debets and Huub Savelkoul, 1999*
98. **Forensic DNA Profiling Protocols,** edited by *Patrick J. Lincoln and James M. Thomson, 1998*
97. **Molecular Embryology:** *Methods and Protocols,* edited by *Paul T. Sharpe and Ivor Mason, 1999*
96. **Adhesion Protein Protocols,** edited by *Elisabetta Dejana and Monica Corada, 1999*
95. **DNA Topoisomerases Protocols:** *II. Enzymology and Drugs,* edited by *Mary-Ann Bjornsti and Neil Osheroff, 1999*
94. **DNA Topoisomerases Protocols:** *I. DNA Topology and Enzymes,* edited by *Mary-Ann Bjornsti and Neil Osheroff, 1999*
93. **Protein Phosphatase Protocols,** edited by *John W. Ludlow, 1998*
92. **PCR in Bioanalysis,** edited by *Stephen J. Meltzer, 1998*
91. **Flow Cytometry Protocols,** edited by *Mark J. Jaroszeski, Richard Heller, and Richard Gilbert, 1998*
90. **Drug–DNA Interaction Protocols,** edited by *Keith R. Fox, 1998*
89. **Retinoid Protocols,** edited by *Christopher Redfern, 1998*
88. **Protein Targeting Protocols,** edited by *Roger A. Clegg, 1998*
87. **Combinatorial Peptide Library Protocols,** edited by *Shmuel Cabilly, 1998*
86. **RNA Isolation and Characterization Protocols,** edited by *Ralph Rapley and David L. Manning, 1998*
85. **Differential Display Methods and Protocols,** edited by *Peng Liang and Arthur B. Pardee, 1997*
84. **Transmembrane Signaling Protocols,** edited by *Dafna Bar-Sagi, 1998*
83. **Receptor Signal Transduction Protocols,** edited by *R. A. John Challiss, 1997*
82. **Arabidopsis Protocols,** edited by *José M Martinez-Zapater and Julio Salinas, 1998*
81. **Plant Virology Protocols:** *From Virus Isolation to Transgenic Resistance,* edited by *Gary D. Foster and Sally Taylor, 1998*
80. **Immunochemical Protocols (2nd. ed.),** edited by *John Pound, 1998*
79. **Polyamine Protocols,** edited by *David M. L. Morgan, 1998*
78. **Antibacterial Peptide Protocols,** edited by *William M. Shafer, 1997*

METHODS IN MOLECULAR BIOLOGY™

Flavoprotein Protocols

Edited by

Stephen K. Chapman

and

Graeme A. Reid

University of Edinburgh, Edinburgh, UK

Humana Press **Totowa, New Jersey**

999 Riverview Drive, Suite 208
Totowa, New Jersey 07512

Cover design by Patricia F. Cleary

This publication is printed on acid-free paper. ∞
ANSI Z39.48-1984 (American Standards Institute)
Permanence of Paper for Printed Library Materials.

For additional copies, pricing for bulk purchases, and/or information about other Humana titles, contact Humana at the above address or at any of the following numbers: Tel.: 973-256-1699; Fax: 973-256-8341; E-mail: humana@humanapr.com; or visit our Website: http://humanapress.com

Printed in the United States of America. 10 9 8 7 6 5 4 3 2 1

Library of Congress Cataloging in Publication Data

Main entry under title:

Methods in molecular biology™.

Flavoprotein protocols/edited by Stephen K. Chapman and Graeme A. Reid.
p. cm.—(Methods in molecular biology; 131)
Includes index.
ISBN 0-89603-734-7 (alk. paper)
1. Flavoproteins—Laboratory manuals. I. Chapman, Stephen K. II. Reid, Graeme A. III. Series: Methods in molecular biology (Totowa, NJ); 131.
QP552.F54F57 1999
572'.68—dc21
DNLM/DLC
for Library of Congress

99-11444
CIP

Preface

As a scientist with an interest in proteins you will, at some time in your career, isolate an enzyme that turns out to be yellow—or perhaps you already have. Alternatively, you may identify a polypeptide sequence that is related to known flavin-containing proteins. This may, or may not, be your first encounter with flavoproteins. However, even if you are an old hand in the field, you may not have exploited the full range of experimental approaches applicable to the study of flavoproteins. We hope that *Flavoprotein Protocols* will encourage you to do so. In this volume we have sought to bring together a range of experimental methods of value to researchers with an interest in flavoproteins, whether or not these researchers have experience in this area.

A broad range of techniques, from the everyday to the more specialized, is described by scientists who are experts in their fields and who have extensive practical experience with flavoproteins. The wide range of approaches, from wet chemistry to dry computation, has, as a consequence, demanded a range of formats. Where appropriate (particularly for analytical methods) the protocol described is laid out in easy-to-follow steps. In other cases (e.g., the more advanced spectroscopies and computational methods) it is far more apt to describe the general approach and relevance of the methods. We hope this wide-ranging approach will sow the seeds of many future collaborations between laboratories and further our knowledge and understanding of how flavoproteins work.

The most common methods are described in the earlier chapters. Anyone who isolates a flavoprotein will almost certainly use UV-visible spectroscopy as the first method of analysis. This simple, yet immensely informative, first approach is concisely described by Macheroux in Chapter 1. Now you have your flavoprotein—does it contain FAD or FMN and how much flavin is present? These questions can be answered by following the methods described by Aliverti et al. in Chapter 2. The electronic structure of flavins makes them amenable to fluorescence-based studies. The fluorimetric methods described in Chapter 3 by Munro and Noble can give useful information on how the protein environment influences the properties of the flavin molecule.

Most flavoproteins are redox active and one of the most important pieces of information relating to their redox activity is the reduction potential of the flavin. In many cases, but not all, it is possible to determine the potentials for both the oxidized-semiquinone and semiquinone-reduced couples. The way to do this is explained clearly by Mayhew in Chapter 4.

A large proportion of flavoproteins, though certainly not all, are enzymes. Chapter 5, by van Berkel et al., describes spectroscopically based methods that allow a detailed analysis of the reaction kinetics of flavoenzymes. Both steady-state and stopped-flow methods are covered, enabling multiple steps in each reaction cycle to be studied. This chapter is complemented by Chapter 6, in which Ingledew describes freeze–quench methods for analyzing reaction intermediates of flavoenzymes, for example, by electron paramagnetic resonance (EPR) spectroscopy. The EPR method is itself described in Chapter 7 by Murataliev. EPR is ideal for observations of the flavin semiquinone, which is stabilized in some proteins and exists as a transient intermediate in the reactions of others.

Circular dichroism is a well-established method for looking at structural aspects of proteins. This technique is of particular use in the study of flavoproteins since the CD spectrum in the region where the flavin absorbs gives valuable information on the environment of the cofactor. Applying the CD technique to flavoproteins is outlined in Chapter 8 by Price et al. Vibrational spectroscopy, particularly surface-enhanced resonance Raman scattering (SERRS), is also of considerable use in the identification of structural and electronic factors associated with the catalytic mechanisms of many flavoenzymes. A protocol for the use of SERRS in such studies is covered in some detail by Macdonald in Chapter 9. Nuclear magnetic resonance, NMR, is a widely used method in the study of many types of protein, but in Chapter 10 Vervoort provides a guide to its particular usefulness in the study of flavin-containing proteins, in which it can provide extensive and detailed information on the flavin environment.

In most flavoproteins the cofactor is noncovalently bound to the protein and can be dissociated. In Chapter 11 Lederer et al. succinctly describe methods for the removal and re-incorporation of flavin. This method is complemented by Chapter 12 in which Edmondson and Ghisla review the use of labeled and chemically altered flavin analogs that can be used to probe the structure and function of the flavin-active site. Reconstituted proteins produced using a combination of procedures from Chapters 11 and 12 can then be analyzed by the spectroscopic and other methods described in the preceding chapters. In some proteins the flavin is covalently attached to an amino acid side-chain, generally altering its functional properties. It is very important to determine the chemical nature of this attachment in order to understand its effect on the properties of the cofactor. In Chapter 13 Scrutton clearly describes methods to determine both the existence and the nature of such a covalent link.

Naturally occurring flavoproteins can be rather complex, and Dutton and colleagues have pioneered new and exciting synthetic approaches to develop simpler models for studying redox–protein function. Chapter 14 describes several methods for constructing such models by attaching flavins to synthetic peptides.

Computational methods are now impinging on almost every aspect of biochemistry, and the study of flavoproteins is no exception. In Chapter 15 Rietjens et al. describe some of the computational methods now available for detailed biochemical analysis of the reactions catalyzed by flavoenzymes.

Finally, Chapter 16 touches on a most important aspect of flavins and flavoenzymes, i.e., their potential for use in medicine. Becker et al. describe a range of diagnostic and therapeutic applications for flavins and flavoproteins that go far beyond the use of riboflavin as a vitamin supplement.

In bringing together the wealth of experience shared by the authors of the individual chapters, we hope that we have provided, in *Flavoprotein Protocols*, a wide-ranging and useful set of methods that can be used by biochemists whether or not they have experience with flavoproteins. It is very unlikely that anyone will use the full range of techniques described here, but by bringing them together in a single volume we hope that you can select those that can be most fruitfully applied to your particular experimental system.

Stephen K. Chapman
Graeme A. Reid

Contents

Contributors

Alessandro Aliverti • *Dipartimento di Fisiologia e Biochimica Generali, Università degli Studi di Milano, Milano, Italy*

Katja Becker • *Center of Biochemistry, University of Heidelberg, Heidelberg, Germany*

Jacques A. E. Benen • *Department of Biomolecular Sciences, Wageningen Agricultural University, Wageningen, The Netherlands*

Bruno Curti • *Dipartimento di Fisiologia e Biochimica Generali, Università degli Studi di Milano, Milano, Italy*

P. Leslie Dutton • *Department of Biochemistry and Biophysics, University of Pennsylvania Medical School, Philadelphia, PA*

Dale Edmondson • *Departments of Biochemistry and Chemistry, Emory University School of Medicine, Atlanta, GA*

Michel H. M. Eppink • *Department of Biomolecular Sciences, Wageningen Agricultural University, Wageningen, The Netherlands*

Garrit Flesichmann • *Laboratory of Enzymology and Biochemical Structures, Centre National de la Recherche Scientifique, Gif-sur-Yvette Cedex, France*

Marco W. Fraaije • *Department of Biomolecular Sciences, Wageningen Agricultural University, Wageningen, The Netherlands*

Sandro Ghisla • *Departments of Biochemistry and Chemistry, Emory University School of Medicine, Atlanta, GA*

Marco Hefti • *Department of Biomolecular Sciences, Wageningen Agricultural University, Wageningen, The Netherlands*

W. John Ingledew • *School of Biomedical Sciences, University of St. Andrews, St. Andrews, UK*

Stefan Kanzok • *Center of Biochemistry, University of Heidelberg, Heidelberg, Germany*

Sharon M. Kelly • *Department of Biological and Molecular Sciences, University of Stirling, Stirling, UK*

Florence Lederer • *Laboratory of Enzymology and Biochemical Structures, Centre National de la Recherche Scientifique, Gif-sur-Yvette Cedex, France*

IAIN D. G. MACDONALD • *Department of Chemistry, University of Strathclyde, Glasgow, Scotland*

PETER MACHEROUX • *Institute of Plant Sciences, Zurich, Switzerland*

STEPHEN G. MAYHEW • *Department of Biochemistry, University College Dublin, Belfield, Dublin, Ireland*

ANDREW W. MUNRO • *Department of Chemistry, University of Edinburgh, Edinburgh, UK*

MARAT B. MURATALIEV • *Department of Entomology, University of Arizona, Tucson, Arizona*

MICHAEL A. NOBLE • *Department of Chemistry, University of Edinburgh, Edinburgh, UK*

NICHOLAS C. PRICE • *Department of Biological and Molecular Sciences, University of Stirling, Stirling, UK*

LARS RIDDER • *Department of Biomolecular Sciences, Wageningen Agricultural University, Wageningen, The Netherlands*

IVONNE M. C. M. RIETJENS • *Department of Biomolecular Sciences, Wageningen Agricultural University, Wageningen, The Netherlands*

HEINZ RÜTERJANS • *Laboratory of Enzymology and Biochemical Structures, Centre National de la Recherche Scientifique, Gif-sur-Yvette Cedex, France*

MARKUS SCHIRMER • *Center of Biochemistry, University of Heidelberg, Heidelberg, Germany*

R. HEINER SCHIRMER • *Center of Biochemistry, University of Heidelberg, Heidelberg, Germany*

NIGEL S. SCRUTTON • *Department of Biochemistry, University of Leicester, Leicester, UK*

R. ERYL SHARP • *Department of Biochemistry and Biophysics, University of Pennsylvania Medical School, Philadelphia, PA*

WILLEM J. H. VAN BERKEL • *Department of Biomolecular Sciences, Wageningen Agricultural University, Wageningen, The Netherlands*

MARIA ANTONIETTA VANONI • *Dipartimento di Fisiologia e Biochimica Generali, Università degli Studi di Milano, Milano, Italy*

JACQUES VERVOORT • *Department of Biomolecular Sciences, Wageningen Agricultural University, Wageningen, The Netherlands*

HANS ZUILHOF • *Department of Biomolecular Sciences, Wageningen Agricultural University, Wageningen, The Netherlands*

1

UV-Visible Spectroscopy as a Tool to Study Flavoproteins

Peter Macheroux

1. Introduction

The essential cofactor of flavoproteins is derived from riboflavin (**Fig. 1**), which consists of the isoalloxazine ring system with a ribityl side chain attached to the central *N*-10 position in the pyrazine moiety.

This precursor is phosphorylated at the 5'-hydroxyl group by flavokinase to yield flavin mononucleotide (FMN). In a second ATP-dependent reaction, FAD pyrophosphorylase attaches an AMP moiety to FMN yielding flavin adenine dinucleotide (FAD). Both cofactors are widely distributed in nature. The side chain plays an important role in binding of the flavin ring system to the protein and therefore the majority of flavoproteins have a strong preference for either FMN or FAD. On the other hand, riboflavin does not play a general role as a cofactor in flavoenzymes. In addition to the noncovalent binding of the flavin to the protein, covalent linkage to an amino acid residue via the 8- or 6-position of the isoalloxazine ring system is also found (*see* Chapter 13). In all cases, the mechanistically relevant moiety is the isoalloxazine ring system that serves as a reversible reduction-oxidation catalyst in many biological redox processes (for a review see: ***(1,2)***). This is also the structural component of the flavin cofactor that is responsible for light absorption in the UV and visible spectral range giving rise to the yellow appearance of the flavin and of flavoproteins (Latin: *flavus* = yellow). In the oxidized form of the flavin two peaks at ≈ 360 and ≈ 450 nm are observed (*see* **Fig. 2**). Method 1 describes a procedure to determine the extinction coefficient for a flavin bound to a particular flavoprotein which is based on the known extinction coefficient for free FMN and FAD (12,500 and 11,300 M^{-1} cm^{-1}, respectively). Binding of the flavin to the protein moiety alters the extinction coefficient covering a range from

From: Methods in Molecular Biology, vol. 131: Flavoprotein Protocols
Edited by: S. K. Chapman and G. A. Reid © Humana Press Inc., Totowa, NJ

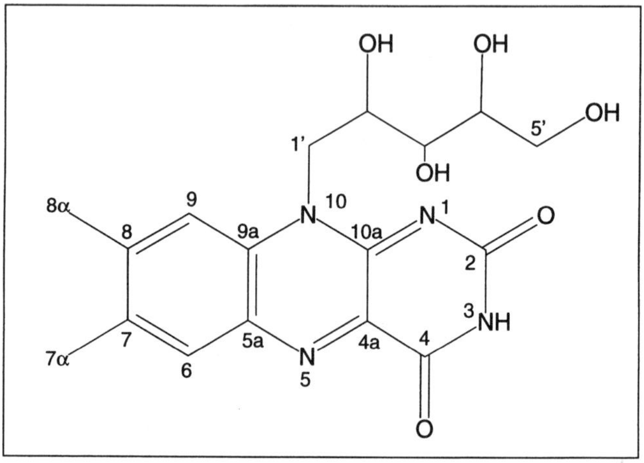

Fig. 1. Structure of riboflavin.

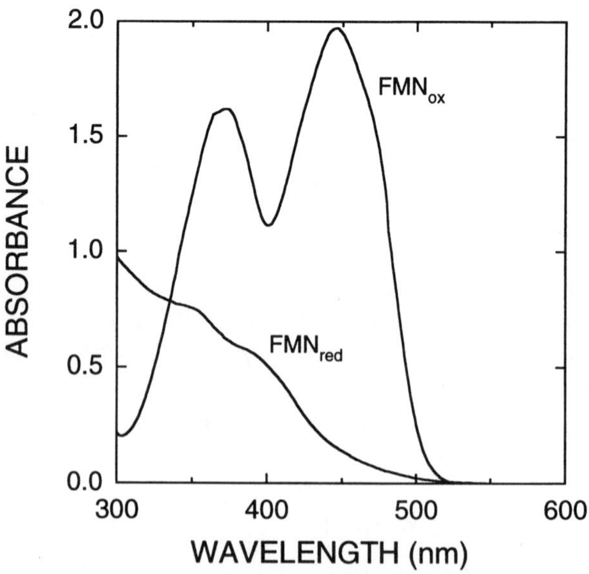

Fig. 2. UV-visible spectrum of FMN in its oxidized and reduced form.

10,500–15,400 M^{-1} cm^{-1} ***(3)***. In some cases the protein lowers the p*K*a for the N(3)-hydrogen (= 10.3 for free flavin) resulting in deprotonation and a lower extinction coefficient. This was observed for glycolate oxidase (9200 M^{-1} cm^{-1}, ***(4)***). Extinction coefficients in that range may therefore indicate that the protein preferentially binds the N(3)-deprotonated flavin.

Knowledge of the extinction coefficient can be helpful to assess the stoichiometry of the flavin and protein moiety. In general one flavin per enzyme active site is found and substoichiometric concentrations of flavin may indicate depletion of the flavin during purification and handling. In these cases it may be possible to reconstitute the flavoprotein by supplementing the flavin cofactor (*see* Chapter 11).

The absorbance in the UV-visible spectral range presents several convenient advantages:

1. Since the flavin chromophore is an essential component of the catalytic machinery, observable spectral changes during the enzymatic reaction yield important information concerning the reaction mechanism. These spectral changes can be observed using rapid reaction kinetic instruments (*see* Chapter 5).
2. Binding of substrates, substrate analogs and inhibitors can be monitored in many cases by spectrophotometry. Titrations with the compound of interest yield dissociation constants and information concerning changes in the active site environment.
3. The protein moiety perturbs the flavin spectrum yielding information on the flavin environment in the binding pocket as well as the ionization state of the flavin.
4. During protein purification, the flavin absorbance can be utilized to locate the fraction of interest (in addition to other parameters, such as activity, if known). This is particularly useful when flavoproteins are prepared from an overproducing host system.

Although some flavoproteins have quite distinctive UV-visible spectra it is not feasible to identify a flavoprotein by the spectral properties of its flavin cofactor. On the other hand, the course of reduction of the oxidized flavin very much depends on the protein moiety: The flavin is either reduced directly to its two-electron reduced form ("hydroquinone") or via its one electron reduced state ("semiquinone"). In the latter case, the flavin semiquinone may either be protonated (neutral or blue semiquinone) or monodeprotonated (anionic or red semiquinone). These two radical species are easily distinguishable by their spectral properties: The neutral semiquinone radical has a characteristic long wavelength absorbance at 580–620 nm ($\varepsilon \approx 4{,}000\ M^{-1}\ cm^{-1}$) whereas the anionic semiquinone radical has a strong absorbance around 380 nm ($\varepsilon \approx 16{,}000\ M^{-1}\ cm^{-1}$) with an additional sharper peak at 400 nm *(5,6)*. It was also recognized that the propensity to stabilize the flavin radical is linked to the enzymatic function of the flavoprotein. This feature was employed to classify the large number of flavoproteins in an attempt to correlate the function of the protein with the properties of its flavin cofactor *(6,7)*. The occurrence of the flavin radical does not only depend on the protein moiety but is also affected by other external physico-chemical factors, such as pH. Equally important is the effect of ligands (substrate, substrate analogs or inhibitors) on the generation of a flavin radical species. Therefore, the reduction of a given and perhaps unknown flavoprotein under various conditions yields important basic information on the type and possible reaction mode of the flavoprotein.

Subheading 2.2, Method 2 provides information on the reducing behavior of a flavoprotein. Reduction of the flavin to its two-electron reduced form results in the loss of the two major absorbance peaks in the UV-visible range and causes bleaching of the flavin chromophore (*see* **Fig. 2**). The procedure described in Method 2 leads to a stepwise reduction of the flavin and hence the absorbance peaks will diminish until complete reduction is achieved with the final spectrum resembling that shown in **Fig. 2** for reduced FMN (FMN_{red}). Another notable feature of such a straightforward reduction to the two-electron reduced flavin is an isosbestic point around 330 nm. However, if a flavin semiquinone is stabilized by the protein moiety then reduction will result in the appearance of a flavin radical identifiable by the spectral characteristics mentioned above *(5,8)*. Consequently, the isosbestic points will change during the course of the photoreduction and will depend on the relative concentrations of oxidized, radical and reduced flavin which are determined by the equilibrium constants of the three redox states. The extent of semiquinone stabilization can be estimated by plotting the absorbance at a wavelength solely due to the semiquinone species (e.g., the long wavelength absorbance of the neutral semiquinone above 580 nm) vs the time of light irradiation (or equivalents of another reductant added) *(9)*. The method described here for reduction of a flavoprotein enjoys a long, successful history and serves as an ideal starting point **(8)**. Further explorations into the reducing properties of a flavoprotein can be achieved with slight alterations of the procedure (*see* **Note 4.**) and with the same equipment (*see* **Subheading 2.2.**).

2. Materials

2.1. Method 1. Determination of the Flavin Extinction Coefficient

1. Spectrophotometer
2. A matched pair of quartz cuvets (1-cm path length, semimicro: 1.2 mL max. vol.)
3. Flavoprotein in a suitable buffer (transparent in the 300–800 nm range)
4. Trichloroacetic acid (50%, v/v)
5. Sodium carbonate (solid)

For the alternative methods discussed in **Subheading 3.1.1.** the following reagents are required:

6. 8 *M* urea or guanidinium hydrochloride
7. Sodium dodecylsulfate (2% w,v)

2.2. Method 2. Photochemical Reduction of Flavoproteins

1. Spectrophotometer
2. Cuvets for anaerobic work (1-cm path length, semimicro: 1.2 mL max. vol.). These cuvets have an upward extension with a stopcock at the outer end in order to connect the cuvet to a gas exchange line. The extension can also serve to equip the cuvet with a side arm in order to deposit other reagents before anaerobicity is established. The same cuvet can be utilized in a glove box (*see* **Subheading 2.2.3.**).
3. Gas exchange line: device that serves to repeatedly evacuate the cuvet with a vacuum pump and exchange the air in the cuvet with nitrogen or argon. Alternatively, a glove box kept under nitrogen or argon can be utilized to achieve anaerobicity. In this case all reagents can be sparkled with nitrogen or argon and then transferred to the glove box. The cuvet can be loaded with the sample (if necessary the side arm is used to deposit a reagent, such as a substrate) and sealed by closing the stopcock.
4. Light source (slide projector, microscope illumination system)
5. Flavoprotein in a suitable buffer containing either 1 m*M* EDTA or potassium oxalate as photosubstrates.

3. Methods

3.1. Method 1. Determination of the Flavin Extinction Coefficient

1. Dissolve 0.5–1 mg of the flavoprotein in an appropriate buffer to give a final volume of 900 µL. For a protein with a molecular mass of $M_r = 50,000$ this will give an approximate absorbance at the maximum at ≈ 450 nm of 0.1–0.2 (depending on the extinction coefficient).
2. Transfer sample to a semimicro cuvet and record an UV-visible absorbance spectrum from 650–300 nm. Read absorbance at the maximum ≈ 450 (reading1)
3. Transfer sample to an Eppendorf tube, add 100 µL of 50% trichloroacetic acid (TCA), spin down white precipitate in a microcentrifuge at maximal speed for 2 min (*see* **Note 2**).
4. Adjust pH to the original pH using solid sodium carbonate.
5. Transfer supernatant (yellow) to semimicro cuvet and record UV/visible absorbance spectrum from 650–300 nm. Correct the obtained spectrum for the dilution factor. Read absorbance at the absorbance maximum at ≈ 450 nm (reading 2). The extinction coefficient of the flavoprotein can now be calculated using the following relationship:

$$\varepsilon_{\text{unknown}} = \varepsilon_{\text{FMN/FAD}} \times \text{reading 1/reading 2}$$

3.2. Method 2. Photochemical Reduction of Flavoproteins

1. Dissolve 0.5–1 mg of the flavoprotein in 1 mL of an appropriate buffer including 1 m*M* EDTA or potassium oxalate and transfer to a cuvet. Record spectrum from 800–300 nm.

2. Remove oxygen with the aid of either a gas exchange line or in a glove box (for a brief description *see* **Subheading 2.2.**). Take care to keep the cuvet sealed until (**step 5**).
3. Record absorbance spectrum in the same range as in (**step 1**) (using the overlay option of the spectrophotometer software).
4. Start reduction by light irradiation. Initially, short intervals of 5–10 s are recommended. Record spectrum after each reduction step until no further spectral changes are observed.
5. Open cuvet to air and record spectrum after complete reoxidation of the sample.

4. Notes

1. In some cases the flavin is covalently linked to the protein moiety and therefore coprecipitates in the denaturation step. However, coprecipitation was also reported for other cases where the flavin is not covalently attached to the protein. Unfolding of the protein in 8 *M* urea or guanidinium hydrochloride and subsequent precipitation of the protein can be helpful to distinguish these two possibilities ***(10)***.
2. Although precipitation of the protein in 5% TCA works well in general, the low pH may also cause decomposition of the flavin. Therefore the time at low pH should be kept to a minimum (*see* **steps 2–4**). In some cases cooling to 4°C helps to prevent such unwanted decomposition. In order to confirm the obtained result, TCA can be replaced by another denaturant, preferably sodium dodecyl sulfate (use 0.2% final concentration).
3. The spectrum recorded in **Subheading 3.2., step 3** should be identical to the one recorded initially. Changes may be observed if the concentration of the flavoprotein is slightly increased due to partial evaporation of the solvent. Cooling the sample with ice during anerobization may help to prevent this. Also, denaturation of the protein may occur during this process (turbidity). This can be spotted at longer wavelength as a baseline drift.
4. Any white light source can be used for irradiation (*see* **Subheading 2.**). For longer irradiation times (>1 min), it is recommended that the sample is shielded from the developing heat of the light source and that the sample be properly thermostatted to the temperature set in the spectrophotometer.
5. Although reduction with light is very convenient for most applications, a chemical reducing agent such as an anaerobic solution of dithionite or an enzymatic system can be used to reduce the flavoprotein. Likewise, where appropriate, a reducing substrate of the flavoenzyme (oxidases, dehydrogenases) can be used to achieve reduction.
6. In some cases photoreduction is slow and the use of µ*M* concentrations of 5-deazariboflavin as a catalyst is recommended ***(11,12)***.
7. Especially when the anionic semiquinone is encountered, further photoreduction to the fully reduced flavoprotein is sometimes difficult to achieve. This situation can be circumvented by using an alternative reductant (*see* above).

References

1. Massey, V. and Ghisla, S. (1983) in *Biological Oxidations* Vol. **34,** pp. 114–139, Springer-Verlag, Mosbach, Germany.
2. Ghisla, S. and Massey, V. (1989) Mechanisms of flavoprotein-catalyzed reactions. *Eur. J. Biochem.* **181,** 1–17.
3. Ghisla, S., Massey, V., Lhoste, J.-M., and Mayhew, S. G. (1974) fluorescence and optical characteristics of reduced flavines and flavoproteins. *Biochemistry* **13(3),** 589–597.
4. Macheroux, P., Massey, V., Thiele, D. J., and Volokita, M. (1991) Expression of spinach glycolate oxidase in *Saccharomyces cervisiae:* Purification and characterization. *Biochemistry* **30,** 4612–4619.
5. Müller, F., Brüstlein, M., Hemmerich, P., Massey, V., and Walker, W. H. (1972) Light-absorption studies on neutral flavin radicals. *Eur. J. Biochem.* **25(3),** 573–580.
6. Massey, V. and Hemmerich, P. (1980) Active-site probes of flavoproteins. *Biochem. Soc. Trans.* **8,** 246–257.
7. Ghisla, S. and Massey, V. (1986) New flavins for old: artificial flavins as active site probes of flavoproteins. *Biochem. J.* **239,** 1–12.
8. Massey, V. and Palmer, G. H. (1966) On the existence of spectrally distinct classes of flavoprotein semiquinones. A new method for the quantitative production of flavoprotein semiquinones. *Biochemistry* **5,** 3181–3189.
9. Mayhew, S. G. (1971) Studies on flavin binding in flavodoxins. *Biochim. Biophys. Acta* **235,** 289–302.
10. Macheroux, P., Hill, S., Austin, S., Eydmann, T., Jones, T., Kim, S.-O., Poole, R., and Dixon, R. (1998) Electron donation to the flavoprotein NifL, a redox-sensing transcriptional regulator. *Biochem. J.*, **332,** 413–419.
11. Massey, V. and Hemmerich, P. (1977) A photochemical procedure for reduction of oxidation-reduction proteins employing deazariboflavin as catalyst. *J. Biol. Chem.* **252(16),** 5612–5614.
12. Massey, V. and Hemmerich, P. (1978) Photoreduction of flavoproteins and other biological compounds catalyzed by deazaflavins. *Biochemistry* **17(9),** 9–17.

2

Identifying and Quantitating FAD and FMN in Simple and in Iron-Sulfur-Containing Flavoproteins

Alessandro Aliverti, Bruno Curti, and Maria Antonietta Vanoni

1. Introduction

The number of known flavin-containing proteins is steadily increasing thanks to the combination of several factors. Among them the following may be of particular interest: (1) increasing power of separation techniques and of molecular biology tools for overproduction of various proteins, and (2) recognition of putative flavin-dependent proteins through analysis of amino acid sequences deduced from those of (putative) genes discovered through genome-sequencing projects. Particularly interesting is the fact that novel flavin-dependent proteins, which play roles different from electron transport or redox catalysis are being discovered (e.g., gene transcription regulation as in the case of NifL, ***(1)***). On the other hand, the picture can be complicated by the fact that the same protein may harbor one or more flavin nucleotides plus one or more additional cofactors ***(2)***. Among such nonflavin centers iron-sulfur clusters are common.

Essential steps for the characterization of flavin-containing enzymes are, (1) the recognition of the presence of a flavin cofactor; (2) the identification of the flavin cofactor(s), and (3) the determination of the stoichiometry of the bound flavin cofactor(s). Several methods are available in the literature in order to achieve such goals, and each flavinologist is indeed familiar with one or more of them ***(3–6)***. We present here three different procedures for both the identification and the quantitation of protein-bound flavin cofactors: a fluorimetric method, a spectrophotometric method, and a high-performance liquid chromatography (HPLC)-based method. These procedures, in our experience, yield reliable qualitative and quantitative information, while requiring limited

From: Methods in Molecular Biology, vol. 131: Flavoprotein Protocols
Edited by: S. K. Chapman and G. A. Reid © Humana Press Inc., Totowa, NJ

amounts of material, minimum sample preparation, and common laboratory equipment.

2. Materials

1. Flavin adenine dinucleotide (FAD) and flavin mononucleotide (FMN) used to construct standard curves are from Sigma Chemical Co. (St. Louis, MO). Stock solutions (5 m*M*) are prepared in MilliQ (Millipore, Bedford, MA) water and stored at –20°C. Individual batches of flavin nucleotides should be checked for the presence of interfering impurities. If required, FAD and FMN can be purified by the HPLC method described in this chapter. The concentrations of FAD and FMN solutions are determined spectrophotometrically, by recording the absorbance spectrum of 25–100 μ*M* solutions in buffer at pH 7–8. The following extinction coefficients are used:

Flavin	λ_{max}	$\varepsilon(M^{-1}cm^{-1})$	Reference
FAD	450	11,300	*7*
FMN	446	12,200	*8*
FAD or FMN	473	9200	This work

2. Sodium dodecyl sulfate (SDS) is from Sigma. A 10% (w/v) solution is prepared in MilliQ water and stored at room temperature. Snake venom phosphodiesterase (PDE) (3 mU/μL) is from Boehringer-Mannheim (Mannheim, Germany), and was stored at 5°C.
3. Enzymes used to test the methods presented are prepared in our laboratories according to standard procedures.
   a. The G298A mutant of glutamate synthase (GltS) β subunit is prepared with a modification of the protocol used to prepare the wild-type recombinant GltS β subunit (P. Morandi, B. Valzasina, and M. A. Vanoni, unpublished data, *see* ref. ***9***), and is stored in 25 m*M* HEPES/KOH buffer, pH 7.5, 10% glycerol at –80°C.
   b. The recombinant GltS α subunit is prepared as described in ***(10)*** and stored in 25 m*M* PIPES/KOH buffer, pH 7.5, 10% glycerol, 0.1 m*M* dithiothreitol (DTT).
   c. The recombinant GltS holoenzyme is prepared using a procedure similar to that used for the preparation of GltS from *Azospirillum* cells (H. Stabile, and M. A. Vanoni, unpublished data, *see* ref. ***11***). It is stored in 25 m*M* HEPES/KOH, pH 7.5, 10% glycerol, 1 m*M* ethylenediaminetetraacetic acid (EDTA), 1 m*M* 2-oxoglutarate and 1 m*M* DTT.
   d. Recombinant spinach leaf ferredoxin-nicotinamide-adenine-dinucleotide phosphate $(NADP)^+$-reductase (FNR) is prepared as described in ***(12)***.
   e. Maize root FNR is produced by heterologous expression in *E. coli* using a modified form of the expression plasmid described in ***(13)***, and purified using a procedure similar to that used for the spinach FNR, replacing anion-exchange with hydrophobic-interaction chromatography on Phenyl-Sepharose (Pharmacia, Uppsala, Sweden) (A. Aliverti et al., manuscript in preparation). Fd/FNR chimeric protein was purified as described ***(14)***.

4. Protein concentration is determined using the method of Bradford ***(15)***, the Amresco Protein Assay Reagent (Amresco, Inc., Solon, OH), and bovine serum albumin as the standard protein. One to 3 μg protein are used for each assay.
5. 10 m*M* HEPES/NaOH, pH 7.5.
6. 10 m*M* Tris/HCl, pH 7.5–8 (at 25°C).
7. 10 m*M* HEPES/KOH, pH 7.5, or 50 m*M* Tris-HCl, pH 7.6 (at 25°C). All buffers (items 5–7) are freshly made in MilliQ water (Millipore) and filtered through 0.2 μm sterile filters.
8. Protein samples are used directly for analyses or after gel filtration through Sephadex G25 (medium) columns. PD10, prepacked disposable columns from Pharmacia are used. All samples are either centrifuged in a microfuge in the cold (top speed, 10 min) or filtered through 0.45 μm filter cartridges or 0.1 μm centrifugal filters (Ultrafree-MC filter units, Millipore) to remove even faint turbidity.
9. Emission spectra are recorded with a Jasco FP-777 spectrophotofluorimeter (Jasco, Inc., Easton, MD) at 20°C.

Fluorimeter settings are as follows:			
λ_{ex}:	450 nm	Excitation slit	5 nm
λ_{em}:	480–600 nm	Emission slit	5 nm
Gain:	High		
Scan speed	100 nm/min		
Response:	1 sec	Data interval	0.5 nm
Instrument zeroing conditions:			
Emission shutter: closed		λ_{ex}: 450 nm	λ_{em}: 480–600 nm

10. Absorbance spectra are recorded with Hewlett-Packard 8453 diode-array (Hewlett-Packard, Palo Alto, CA), Cary 219 (Varian, Palo Alto, CA) or Uvikon 810 (Kontron Elektronik, Eching, Germany) spectrophotometers at 20 or 25°C, with identical results.
11. HPLC analyses are carried out using a μBondapack C-18 silica column (Waters Associates, Milford, MA) connected to a Waters 600E HPLC system equipped with a UK6 manual injector and controlled through the Millenium software (Waters) run on a PC. A 2-mL sample loop is used, and the column is not thermostated. Eluate absorbance is monitored continuously using an on-line 486 detector (Waters) set at 264 nm. Solvents are 5 m*M* ammonium acetate, pH 6.5, and methanol (HPLC grade). Both solvents are filtered through 0.22 μm filters and degassed by continuous helium bubbling.

3. Methods

3.1. Fluorimetric Identification of the Flavin Cofactor as FAD or FMN

The method of Forti and Sturani ***(16)*** is used. It exploits the fact that fluorescence of a FMN solution is 10-fold higher than that of an FAD solution of the same concentration, and that phosphodiesterase (PDE) catalyses the conversion of FAD into FMN and adenosine monophosphate (AMP). Thus, the fluorescence intensity of a flavin-containing solution is measured before and after addition of PDE. No increase of fluorescence is interpreted as due to absence of FAD in the solution; a 10-fold increase of fluorescence allows the identification of the flavin present as FAD, an intermediate increase of fluorescence allows the calculation of the relative concentration of FAD and FMN in the starting solution. Several modifications of the following method are possible, depending on the quantity of enzyme and on the equipment available. This method can also be employed to test the purity of the FAD stock solution used.

3.1.1. Calibration and Linearity Test of the Assay

It is important to determine the range of linearity of the instrument by measuring the emission spectrum of solutions containing increasing concentrations of FAD or FMN in the same buffer, in the same cuvet, and under the same settings that will be used for the actual experiment.

1. A 0.25–0.3 m*M* solution of FAD or FMN is prepared in water, or in the buffer that will be used for the enzyme solution. Its concentration is determined by measuring the absorbance spectrum (*see* **Subheading 2.1.**).
2. One microliter aliquots are added to a cuvet containing 2 mL buffer. Alternatively, a 200 µL cuvet can be used (*see* **Note 1**).
3. After mixing, emission spectra are recorded.
4. The emission intensity at 524 nm, corrected for buffer emission value and dilution (F_{524}), is plotted as a function of flavin concentration.
5. The initial linear part of the curve sets the interval of fluorescence values that will yield reliable results. From the slope of the line $F_{524} = f_{FAD}[FAD]$ or $F_{524} = f_{FMN}[FMN]$, the values of intensity of light emitted at 524 nm by a 1 µ*M* solution of FAD (f_{FAD}) or FMN (f_{FMN}) are also calculated.

3.1.2. Fluorimetric Analysis of the Cofactor(s) Released From the Flavoprotein Solution

1. A 5–10 µ*M* solution of enzyme is prepared by gel filtration through a Sephadex G25 (Pharmacia) column equilibrated with either 10 m*M* HEPES/NaOH buffer, pH 7.5 or 10 m*M* Tris/HCl buffer, pH 7.5.
2. The absorbance spectrum is recorded, and the protein and activity content of the sample is measured.
3. The emission spectrum of 2 mL buffer is measured in a 3 mL glass cuvet to be used as a blank in subsequent measurements.
4. A 100 µL aliquot of the enzyme solution is added to the fluorimeter cuvet.
5. After mixing, the emission spectrum of the solution is recorded.
6. The sample is recovered with a Pasteur pipet, transferred to microfuge tubes, wrapped in aluminum foil, and incubated for 10 min at 100°C.

7. After being cooled on ice, microfuge tubes are centrifuged at 4°C for 10 min at 13,000 rpm (14,500*g*).
8. The supernatant is recovered and transferred directly into the fluorimeter cuvet.
9. The emission spectrum of this solution is measured as before.
10. 2 µL (6 mU) of the PDE solution are added.
11. The emission spectrum of the solution is recorded at different times after mixing, until no further changes are observed.
12. Both the emission and excitation shutter of the instrument are closed between measurements to avoid photodegradation of the flavin sample. Within 5 min the maximum increase of emission at 524 nm is obtained. After 20–30 min, emission tends to decrease (*see* **Note 2**).
13. Depending on the results obtained, measurements are repeated using (a) a different sample dilution (e.g., 5- or 50-fold dilution), (b) a different incubation time at 100°C (5, 15, or 20 min), and (c) a different sample denaturation method (*see* **Subheading 3.1.3.**). If limited by the amount of sample available, 200 µL cuvets may be used.

3.1.3. Alternative Protein Denaturation Methods

1. The protein stock solution (5–10 µ*M*) can be denatured by incubation at 100°C for 10 min in the dark, and denatured protein can be removed by centrifugation at 13,000 rpm (14,500*g*) for 10 min in a microfuge in the cold. The supernatant is transferred to a clean microfuge tube, kept on ice in the dark and aliquots can be diluted 10–50-fold for fluorimetric analyses as described in 3.1.2.
2. The protein stock solution (5–10 µ*M*) can be denatured by incubation at room temperature in the presence of 0.2% SDS (*see* **Subheading 3.2.1.**). Aliquots of such sample can be diluted 10–20-fold for fluorescence analyses described above. Although 0.2% SDS does not allow PDE-catalyzed conversion of FAD into FMN (*see* **Table 1**), we observed that experiments carried out in the presence of 0.02% SDS gave results essentially indistinguishable from those obtained in the absence of SDS.

3.1.4. Fluorimetric Data Analysis

As already reported, if the protein contains only FMN, no fluorescence change is observed upon PDE treatment. Conversely, if the protein contains only FAD a 10-fold fluorescence increase is expected. For proteins containing both flavin cofactors, the molar ratio (*r*) between FAD and FMN can be calculated from the fluorescence increase after PDE treatment using Eq. 4, which is based on Eqs. 1–3.

$$F_o = f_{FAD} \cdot [FAD] + f_{FMN}[FMN] \quad (1)$$

$$F_{fin} = f_{FMN}\,([FMN] + [FAD]) \quad (2)$$

$$f_{FAD} = 0.1 f_{FMN} \quad (3)$$

$$r = (10 \times (F_{fin}/F_o) - 10) \,/\, (10 - (F_{fin}/F_o)) \quad (4)$$

Table 1
Fluorimetric Identification of FAD and FMN in GltS Holoenzyme and in Its Isolated Subunits

Sample	Concentration[a] (μ*M*)	Buffer	Pretreatment	Denaturation method	F_o	F_{fin}	F_{fin}/F_o	Notes
FAD	0.60	A + 0.1 m*M* $MgSO_4$	None	None	20.63	206.3	10	Chromatographically purified FAD
FAD	0.26	A + 0.1 m*M* $MgSO_4$	None	None	39.06	202	5.17	Commercial FAD
GltS β[b]	0.781	B + 0.1 m*M* $MgSO_4$	Gel filtration	10 min, 100°C	56.6	611.8	10.81	
GltS β[b]	0.443	A + 0.1 m*M* $MgSO_4$	Gel filtration	10 min, 100°C	37.6	399.1	10.6	
GltS β[b]	0.762	A	Gel filtration	0.2% SDS to 7.62 μ*M* β subunit stock solution	80.3	777.8	9.72	SDS concentration in fluorimeter cuvet was 0.024%
GltS β[b]	0.381	A	Gel filtration	0.2% SDS to 7.62 μ*M* β subunit stock solution	40.22	374.4	9.3	SDS concentration in fluorimeter cuvet was 0.012%
GltS β[b]	0.381	A + 0.1 m*M* $MgSO_4$	Gel filtration	0.2% SDS to 7.62 μ*M* β subunit stock solution	40.84	410.5	10.5	SDS concentration in fluorimeter cuvet was 0.012%
G298A-GltSβ[c]	9.56	A	Gel filtration	0.2% SDS	32.7	69.94	—	PDE activity inhibited by 0.2% SDS

G298A-GltSβ[b]	0.212	A + 0.1 m*M* $MgSO_4$	Gel filtration	10 min, 100°C	21.35	169	7.916	The enzyme was not shielded from light during heat denatura
G298A-GltSβ[c]	0.429	A + 0.1 m*M* $MgSO_4$	Gel filtration	10 min, 100°C	32.26	316	9.79	The sample was shielded from light during heat denatura
GltS α[c]	0.46	A	Gel filtration	10 min, 100°C	93.3[d]	91.5[d]	0.98	
rGltS[c,e]	4.81	C	None	10 min, 100°C	356.3	645.2	1.81	
rGltS[c,e]	0.534	C	None	10 min, 100°C	39.65	82.47	2.07	same sample as above, diluted 9-folc

[a]Protein concentration was determined by the method of Bradford ***(15)***.
[b]In 3 mL cuvet containing 2220 μL sample.
[c]In 200 μL cuvet with 200 μL sample volume.
[d]Excitation light was at 440 nm.
[e]In 25 m*M* Hepes/KOH, pH 7.5, 1 m*M* EDTA, 1 m*M* DTT, 1 m*M* 2-oxoglutarate, 10% glycerol,
Buffers were A, 10 m*M* Tris/HCl, pH 7.5; B, 10 m*M* HEPES/NaOH, pH 7.5; C, 10 m*M* HEPES/KOH, pH 7.5.

where F_o = fluorescence of the released flavin solution; F_{fin} = fluorescence after reaction with PDE; f_{FAD} = fluorescence of a 1 μ*M* solution of FAD; and f_{FMN} = fluorescence of a 1 μ*M* solution of FMN. Fluorescence refers to the intensity of light emitted at 524 nm on excitation with 450 nm light.

In the case of the GltS β subunit or of its G298A mutant, from the changes of emission intensity of the solutions (**Table 1**) it can be concluded that the enzyme contains FAD as the flavin cofactor. The results are not influenced by the presence of 0.1 m*M* $MgSO_4$, which was called for in the original method *(16)*, nor by the incubation time at 100°C. Instead, we found it important to protect the sample from light throughout the treatment. As shown in **Table 1**, no fluorescence changes were obtained on incubation of the flavin cofactor released from a sample of GltS α subunit indicating that the protein contains FMN as the flavin cofactor beside a 3Fe/4S cluster *(10)*. Fluorescence changes on addition of PDE to the flavins released from GltS holoenzyme are more complex. The increase of fluorescence is about twofold, consistent with the presence of approximately equimolar amounts of FAD and FMN in the GltS holoenzyme (**Table 1**). A more precise estimate of the flavin stoichiometry is obtained by HPLC analysis of the released flavins (*see* **Subheading 3.3.**).

The method can be used to calculate the FAD and/or FMN concentration of the solution under analysis provided the intensity of emission at 524 nm as a function of FAD or FMN concentration has been determined experimentally. Under our conditions, using GltS, the emission intensity of solutions containing between 1 and 8 μ*M* FMN was measured in the 200 μL cuvet, so that it was determined that a 1 μ*M* FMN solution yields a fluorescence intensity of 76.4. Assuming that the GltS protomer contains one FAD and one FMN, from the fluorescence after the PDE treatment, an enzyme concentration of 4.8 μ*M* could be calculated in agreement with the determination made from protein assays (4.8 μ*M*).

3.2. Spectrophotometric Determination of the Flavin Content, Identification of the Flavin Cofactor as FAD or FMN, and Calculation of Extinction Coefficients

3.2.1. SDS Treatment of Simple Flavoproteins

1. The absorbance spectrum of the protein solution (5–10 μ*M*, 1 mL) in 10 m*M* HEPES/NaOH buffer, pH 7.5, or 10 m*M* Tris-HCl buffer, pH 7.5–8, is recorded.
2. A 20 μL aliquot of a fresh 10% SDS solution is added (*see* **Note 3**).
3. After mixing, spectra are recorded until no further changes are observed. Typically, within 5–10 min conversion of the spectrum of the bound flavin into that of free FAD or FMN is observed.
4. Direct comparison of equimolar solutions of FAD and FMN reveals that at 473 nm the solutions exhibit a similar extinction coefficient (9200 $M^{-1}cm^{-1}$).

Thus, the following conclusions can be drawn from the comparison of the spectrum of the native protein with that obtained after SDS addition:

a. From the position of the absorbance maximum at 450 vs 446 nm, or of the absence of a well-defined peak between 446–450 nm the presence of FAD, FMN or of a mixture of the two in the released flavin solution can be inferred.
b. Once the type of flavin present is identified, its concentration can be precisely estimated using ε_{450} = 11,300 $M^{-1}cm^{-1}$ for FAD and ε_{446} = 12,200 $M^{-1}cm^{-1}$ for FMN. In case of doubt, or of other information, or if a mixture of FAD and FMN may be present, the total (FAD + FMN) concentration can be calculated from the absorbance at 473 nm, using 9200 $M^{-1}cm^{-1}$ as the extinction coefficient.
c. Once the concentration of the flavin is known, both the flavin/protein molar ratio and the extinction coefficient of the bound flavin can be calculated.

3.2.2. Heat Denaturation of Simple and Complex Flavoproteins

SDS-treatment sometimes fails to denature proteins, in particular those containing iron-sulfur centers as in our case (i.e., GltS, the GltS α subunit, the chimeric Fd/FNR fusion protein). An effective method to destroy the protein's iron-sulfur clusters, while preserving the flavin cofactors, is heat treatment in the dark.

1. The spectrum of the protein solution (5–10 μ*M*, 1 mL) is recorded.
2. The sample is transferred to a microfuge tube, and incubated in a boiling water bath for 10 min.
3. After cooling, the denatured protein is removed by centrifugation (10 min at 13,000 rpm (14,500*g*) in the cold in a microfuge).
4. The supernatant is recovered, and the absorbance spectrum is recorded (*see* **Note 4**).
5. If centrifugation is not sufficient to remove denatured protein the supernatant can be either filtered through 0.2 μm filters, or using centrifugal filters.
6. Again the identification of the flavin can be attempted from the position of the absorbance maximum (450 nm for FAD as opposed to 446 nm for FMN).
7. Flavin quantitation can be carried out as described above (*see* **Subheading 3.2.1.**), allowing determination of stoichiometry and extinction coefficient of the bound flavin.

3.2.3. Spectrophotometric Identification of FAD and FMN by PDE Treatment Following Heat Denaturation

The differences of the absorbance spectra of FAD and FMN can be exploited to carry out the identification of FAD and FMN using the PDE treatment, directly in the spectrophotometer cuvet.

1. The spectrum of a sample containing 5–10 μ*M* flavin (1 mL) is recorded.
2. A 2 μL aliquot of PDE is then added, and spectra are recorded at different times after PDE addition.

3. Using a FAD-containing solution, absorbance changes consistent with conversion of FAD into FMN (+ AMP) are observed with an isosbestic point at 473 nm, and they are complete within 10 min. As expected, no changes are observed when a FMN-containing solution is treated with PDE.

3.2.4. Quantitation of Total Flavin Content in Iron-Sulfur Flavoproteins

Determination of the non-heme iron content of iron-sulfur containing proteins is often affected by the precision of the protein concentration determination. In order to avoid this problem with GltS, a method was designed ***(17)*** that allowed us to measure non-heme iron concentration and total flavin concentration on the same sample derived from heat denaturation, under acidic conditions, of GltS or GltS α subunit solutions.

1. 0.8 mL Samples of GltS or GltS α subunit (5–10 μ*M*) that have been gel filtered in 10 m*M* HEPES/KOH buffer, pH 7.5 or 10 m*M* Tris/HCl buffer, pH 7.5–8, are prepared.
2. The absorbance spectrum of the solution is recorded, and the solution is transferred into a microfuge tube. One-tenth microliter of 100% trichloroacetic acid (TCA) and 0.1 mL 75 m*M* ascorbic acid are added.
3. After being mixed, the sample is incubated in a boiling water bath for 5–10 min.
4. The combination of acid and heat treatment results in hydrolysis of FAD present into FMN.
5. Inclusion of ascorbic acid helps in maintaining Fe in the +2 form required for iron concentration determination.
6. After being cooled, denatured protein is removed by centrifugation for 10 min at 13,000 rpm (14,500*g*) in a microfuge in the cold.
7. The absorbance spectrum of the supernatant is measured, and the total flavin concentration is determined using 11.1 $mM^{-1}cm^{-1}$ as the extinction coefficient of FMN at 446 nm, under acidic conditions ***(8)***.
8. The total flavin concentration can be used as an estimate of the GltS concentration, assuming that 1 FAD and 1 FMN are present per enzyme protomer, or that 1 FMN is present per GltS α subunit.
9. Enzyme concentration determined through protein assays and through the total flavin content never differed by more than 20% (**Table 2**). The supernatant from the heat denaturation step under acidic conditions can be directly used for the determination of the iron content of the preparation ***(17)***.

3.3. HPLC Identification and Quantification of FAD and FMN

Several methods for the separation of flavins by reverse phase chromatography have been published. We propose a modification of the method of Light et al. ***(18,19)*** which uses 5 m*M* ammonium acetate, pH 6.5, and methanol as the solvent system. Thus, it allows both to separate FAD and FMN on an analytical scale, and to obtain homogeneous preparations of FAD or FMN from commercially available compounds. When used on a preparative scale, the solvent

Table 2
Spectrophotometric Quantitation of FAD and FMN Extracted from Several Flavin-Containing Enzymes

Enzyme	Concentration[a] (μ*M*)	Pretreatment	Buffer	Denaturation	Native λmax, A@λmax	Denatured λmax, A@λmax	Flavin concentration (μ*M*)	Stoichiometry	ε ($M^{-1}cm^{-1}$)
GltS β subunit	16.56	Gel filtration	C	0.2% SDS	454, 0.157	450, 0.152	(FAD), 13.4	0.81	11,698
GltS β subunit	8.85	Gel filtration	B	0.2% SDS	454, 0.083	450, 0.085	(FAD), 7.5	0.85	11,080
G273A GltS β subunit	8.8	None	A	10 min, 100°C	452, 0.099	450, 0.086	(FAD), 7.6	0.86	12,955
G273A GltS β subunit	4.3	Gel filtration	B	0.2% SDS	452, 0.049	450, 0.048	(FAD), 4.3	0.99	11,448
GltS α subunit	2.3	Gel filtration	B	10 min, 100°C	440, 0.046	446, 0.029	(FMN), 2.4	1.04	19,200
rGltS	4.8	Gel filtration	D	10 min, 100°C	444, 0.300	448, 0.115	(FAD + FMN)		
						473, 0.09[c]	9.8[c]	2.04	61,200
rGltS	3.35	Gel filtration	B	10 min, 100°C[b]	444, 0.210	446, 0.07[d]	6.2[d]	1.85	67,742

[a]Enzyme concentration was determined by the Bradford method ***(15)***.
[b]In the presence of 10% TCA, 7.5 m*M* ascorbic acid, and light.
[c]The extinction coefficient at the isosbestic point for FAD and FMN (9200 $M^{-1}cm^{-1}$) was used.
[d]The extinction coefficient of FMN under acidic conditions is used (11,100 $M^{-1}cm^{-1}$, ***8***).
A, 25 m*M* HEPES/KOH, pH 7.5, 10% glycerol, 1 m*M* EDTA; B, 10 m*M* Tris/HCl, pH 7.5; C, 10 m*M* HEPES/NaOH, pH 7.5; D, 10 m*M* HEPES/KOH, pH 7.5.

can be easily removed by lyophilyzation. The eluate can be monitored continuously by absorbance or fluorescence, depending on the online detection system available. We routinely monitor the absorbance of the eluate at 264 nm, where the absorbance of FAD and FMN solutions are maximal. Greater sensitivity could be obtained by measuring the eluate absorbance at shorter wavelengths, but quantitation may be affected by baseline absorbance changes due to varying eluent composition.

1. A 2 mL sample loop is mounted on the UK-6 injector and varying sample volumes (50–250 µL) are injected onto the column using microsyringes.
2. The column is equilibrated at a flow rate of 1 mL/min for 30 min with 85% solvent A (5 m*M* ammonium acetate buffer, pH 6.5) and 15% methanol.
3. After sample injection the column is washed with 85% solvent A and 15% methanol for 5 min.
4. The concentration of methanol is increased linearly from 15–75% in 20 min and from 75–100% in 5 min.
5. Methanol concentration is kept constant at 100% for 10 min and lowered to 15% in 5 min.

3.3.1. Construction of the Calibration Curve

1. A calibration curve is constructed by injecting 50 µL aliquots of solutions containing known amounts of FAD and FMN (40–800 pmol each).
2. Under these conditions, FAD elutes as a sharp peak at 9.5 min, and FMN as a broader peak at 18 min.
3. Integration of the areas corresponding to the peaks obtained by injecting known amounts of FAD and FMN allows the construction of standard curves.
4. Under these conditions, peak position and quantitation is independent from sample volume (50–250 µL of a 1 µ*M* solution of FAD and FMN). Moreover, incubation of FAD and FMN solutions at 100°C for up to 20 min does not lead to any detectable change of the chromatogram, provided the samples are kept in the dark. Furthermore, freezing of such solutions for up to one week also does not lead to any detectable change. On the contrary, exposure of FAD and FMN solutions to light, to acidic conditions (10% TCA) or to both acidic conditions (10% TCA) and heat (10 min at 100°C) leads to conversion of FAD into FMN, and (presumably) AMP, which is not retained by the column.

3.3.2. Sample Preparation and Analysis of the Released Cofactors

1. Flavoprotein samples are diluted to a final concentration of 5–20 µ*M*.
2. Protein solutions are transferred to microfuge tubes and incubated at 100°C for 10 min.
3. The denatured protein is removed by centrifugation, and the supernatant is filtered.
4. The absorbance spectrum of each solution is measured, and 50, 100, and 150 µL aliquots are injected onto the HPLC.
5. To test the method, samples of GltS holoenzyme, GltS α subunit, spinach leaf FNR, maize root FNR, and the chimeric Fd/FNR protein are analyzed. As shown in **Table 3**, the chromatographic analysis reproducibly yields results consistent

Table 3
HPLC Quantification of the Flavin Content of Several Flavoproteins

Experiment	Enzyme	Concentration[a] (μ*M*)	Buffer[b]	Denaturation method	Enzyme (pmol)	FAD (pmol)	FMN (pmol)	FAD/ subunit	FMN/ subunit
1	rGltS	5.8	A	100°C, 10 min	290	247	248	0.85	0.85
					580	471	507	0.81	0.87
2	GltS α	5.3	B	100°C, 10 min	265	None	275	—	1.04
					530	None	520	—	0.98
3	Leaf-FNR	5.50	C	100°C, 10 min	275	262	None	0.95	—
					550	527	None	0.96	—
4	Root-FNR	4.86	C	100°C, 10 min	243	230	None	0.95	—
					486	469	None	0.97	—
5	Fd/FNR	5.52	C	100°C, 10 min	276	289	None	1.05	—
					552	588	None	1.07	—

[a]Concentration determined with the Bradford method ***(15)***.

[b]Buffers used were: A, 25 m*M* Hepes/KOH, pH 7.5, 10% glycerol, 1 m*M* EDTA, 1 m*M* 2-oxoglutarate, 1 m*M* DTT; B, 25 m*M* Pipes/KOH, pH 7.5, 10% glycerol, 1 m*M* EDTA, 0.1 m*M* DTT; C, 50 m*M* Tris-HCl, pH 7.4.

with the known flavin content of the enzymes, as determined previously and by using the alternative methods presented above (*see* **Subheadings 3.1.** and **3.2.**).

4. Notes

1. When 200 µL cuvets are used, care must be taken to transfer the samples with automatic pipets, and to start from at least 250 µL samples. Mixing must also be carried out with automatic pipets. We discourage the use of microsyringes when the sample contains PDE, as the enzyme is difficult to wash away and may contaminate samples that will be handled later on.
2. We recommend soaking cuvets that have come in contact with PDE in a solution containing 3 *M* HCl in ethanol and extensively rinsing with MilliQ water (Millipore) for cleaning and effective removal of PDE traces.
3. In these experiments, it should be kept in mind that due to the limited solubility of potassium dodecyl sulfate, the protein should be transferred into a potassium-free buffer.
4. Comparison of the spectrum of the flavin released by the SDS treatment and that released by heat denaturation sometimes yields different results: up to 20% flavin loss during heat denaturation has been observed, presumably due to trapping of some of the flavin in the protein pellet. Thus, it is recommended that the experiment is repeated using protein solutions of different concentrations.

Acknowledgments

The work carried out in the authors' laboratory is funded by grants from the Ministero per l'Università e la Ricerca Scientifica and Consiglio Nazionale delle Ricerche (Target Project on Biotechnology). We are grateful to Dr. L. Piubelli for providing FNR samples. The contribution of Drs. F. Fischer, P. Morandi, S. Ravasio, H. Stabile, and B. Valzasina, who provided samples of GltS holoenzyme and of its two subunits and tested the methods described above, is gratefully acknowledged.

References

1. Hill, S., Austin, S., Eydmann, T., Jones, T., and Dixon, R. (1996) *Azotobacter vinelandii* NifL is a flavoprotein that modulates transcriptional activation of nitrogen-fixation genes via a redox sensitive switch. *Proc. Natl. Acad. Sci. (USA)* **93,** 2143–2148.
2. Palfey, B. A. and Massey, V. (1998) Flavin Dependent Enzymes, in *Comprehensive Biological Catalysis* (Sinnott, M. L., ed.) vol. III, Academic Press, New York, pp. 83–154.
3. McCormick, D. B. and Wright L. D., eds. (1971) *Methods in Enzymology* vol. 18, Part B, Academic, London and New York, pp. 217–425.
4. Fleischer, S. and Packer, L., eds. (1978) *Methods in Enzymology* vol. 53, Academic, New York, pp. 397–570.
5. McCormick, D. B. and Wright L. D., eds. (1980) *Methods in Enzymology* vol. 66, Part E, Academic, London and New York, pp. 253–598.

6. Chytil, F. and McCormick, D. B., eds. (1986) *Methods in Enzymology* vol. 122, Academic, London and New York, pp. 185–248.
7. Koziol, J. (1971) Fluorometric analyses of riboflavin and its coenzymes, in *Methods in Enzymology*, (McCormick, D. B. and Wright, L. D., eds.), vol. 18, Part B, Academic, London and New York, pp. 253–285.
8. Hinkson, J. W. (1968) *Azotobacter* free-radical flavoprotein. Preparation and properties of the apoprotein. *Biochemistry* **7,** 2666–2672.
9. Vanoni, M. A., Verzotti, E., Zanetti, G., and Curti, B. (1996) Glutamate synthase: properties of the recombinant β subunit. *Eur. J. Biochem.* **236,** 937–946.
10. Vanoni, M. A., Fischer, F., Ravasio, S., Verzotti, E., Edmondson, D. E., Hagen, W. R., Zanetti, G., and Curti, B. (1998) The recombinant α subunit of glutamate synthase: spectroscopic and catalytic properties. *Biochemistry* **37,** 1828–1838.
11. Ratti, S., Curti, B., Zanetti, G., and Galli, E. (1985) Purification and characterization of glutamate synthase from *Azospirillum brasilense. J. Bacteriol.* **163,** 724–729.
12. Aliverti, A., Bruns, C. M., Pandini, V. E., Karplus, P. A., Vanoni, M. A., Curti, B., and Zanetti, G. (1995) Involvement of Serine 96 in the catalytic mechanism of ferredoxin-NADP$^+$ reductase: structure-function relationship as studied by site-directed mutagenesis and X-ray crystallography. *Biochemistry* **34,** 8371–8379.
13. Ritchie, S. W., Redinbaugh, M. G., Shiraishi, N., Vrba, J. M., and Campbell, W. H. (1994) Identification of a maize root transcript expressed in the primary response to nitrate: characterization of a cDNA with homology to ferredoxin-NADP$^+$ oxidoreductase. *Plant. Mol. Biol.* **26,** 679–690.
14. Aliverti, A. and Zanetti, G. (1997) A three-domain iron-sulfur flavoprotein obtained through gene fusion of ferredoxin and ferredoxin-NADP$^+$ reductase from spinach leaves. *Biochemistry* **36,** 14,771–14,777.
15. Bradford, M. M. (1976) A rapid and sensitive method for the quantitation of microgram quantitites of protein utilizing the principle of protein-dye binding. *Anal. Biochem.* **72,** 248–254.
16. Forti, G. and Sturani E. (1968) On the structure and function of reduced nicotinamide adenine dinucleotide phosphate-cytochrome f of reductase of spinach chloroplasts. *Eur. J. Biochem.* **3,** 461–472.
17. Vanoni, M. A., Edmondson, D. E., Zanetti, G., and Curti, B. (1992) Characterization of the flavins and the iron-sulfur centers of glutamate synthase from *Azospirillum brasilense* by absorption, circular dichroism, and electron paramagnetic resonance spectroscopies. *Biochemistry* **31,** 4613–4623.
18. Light, D. R., Walsh, C., and Marletta M. A. (1980) Analytical and preparative high-performance liquid chromatography separation of flavin and flavin analog coenzymes. *Anal. Biochem.* **109,** 87–93.
19. Hausinger, R. P., Honek, J. F., and Walsh, C. (1986) Separation of flavins and flavin analogs by high-performance liquid chromatography, in *Methods in Enzymology* (Chytil, F. and McCormick, D. B., eds.), vol. 122, Academic, New York, pp. 199–209.

3

Fluorescence Analysis of Flavoproteins

Andrew W. Munro and Michael A. Noble

1. Introduction

Fluorescence is one of the most powerful tools in the study of flavoproteins. It allows the user to investigate many aspects of the nature, structural integrity and kinetic properties of the flavin. In this chapter, we will outline the basics of fluorescence spectroscopy and give details of the operating procedures required to obtain different kinds of information about the protein-bound chromophore. Leading on from this will be a number of examples of the uses of fluorescence in flavoprotein analysis. In **Subheading 4.**, the potential benefits and drawbacks of fluorescence spectroscopy in the study of flavoproteins will be discussed.

1.1. An Introduction to Fluorescence Spectroscopy

A detailed account of processes accompanying the absorption of incident radiation by matter (biological or non-biological) is beyond the scope of this chapter and the reader is advised to consult texts that deal with fundamental theory of electronic excitation and its consequences (e.g., ***1,2*** and references within). However, a basic knowledge of fluorescence will allow the reader to comprehend the basics of the technique and to apply it successfully in the analysis of flavins. A brief explanation of fluorescence theory is given as follows.

Fluorescence spectroscopy, as with conventional absorption spectrophotometry, involves the excitation of a sample with energy from the UV-visible region of the electromagnetic spectrum (approx. 200–700 nm wavelength range). Both techniques rely on the fact that the molecule being examined absorbs photons of energy at defined wavelengths. The absorption of an incident photon occurs when its energy is exactly that required to promote a "transition" of an electron from a lower to a higher energy state ***(3)***. Prior to the absorption of a subsequent photon, the energy must be lost from the molecule and it must return to its

From: Methods in Molecular Biology, vol. 131: Flavoprotein Protocols
Edited by: S. K. Chapman and G. A. Reid © Humana Press Inc., Totowa, NJ

ground state. This is usually a very fast process (taking less than ~ 1 ps [10^{-12} s]), occurring due to vibrations and rotations within the molecule, or through collision with other molecules—particularly solvent. However, for some molecules other means of energy loss are available. These molecules often have rather rigid, conjugated, double-bond systems, and may lose energy via vibration/rotation quite slowly. If this is the case, then radiative emission of a photon may occur. If this emission occurs from a singlet excited state (i.e., one in which the excited electron remains spin-paired with its previous ground state partner), then it is by fluorescence. If it occurs after inter-system crossing to a triplet excited state, then it is by phosphorescence. For fluorescence, the lifetime of the excited state is ~ 1 ns [10^{-9} s]; for phosphorescence the lifetime is ~ 1 ms [10^{-3} s] *(3)*. Thus, the fact that flavins (*see* Chapter 8 for diagrams) fluoresce is indicative of the fact that rotational and vibrational motion in their excited (singlet) states is considerably constrained—allowing significant energy loss as radiation. Not surprisingly, given the competition between radiative and nonradiative forms of energy loss, the energy emitted as fluorescence is less than that absorbed by the molecule i.e., fewer photons are emitted than are absorbed. This ratio is termed the "quantum yield" (Q_f) *(3)*. The Q_f varies between 0 and 1, but is always less than 1 because a lower number of photons are emitted compared to that absorbed. There is also generally nonradiative loss of energy to the environment during the lifetime of the singlet state of the molecule. The consequence of this is that the energy of emitted photons is lower than that of those absorbed—and fluorescence is observed at lower energy (i.e., longer wavelength) than the exciting radiation.

1.2. Flavin Fluorescence

All molecules (biological or otherwise) absorb photons. However, relatively few fluoresce. Two of the most important types of biomolecule that *do* fluoresce (i.e., those which have proved useful in analytical procedures) are: 1. aromatic amino acids, particularly tryptophan, the indole group of which absorbs UV light (λ_{max} = 280 nm, extinction coefficient, ε, = 5050 M^{-1} cm^{-1}) and fluoresces with an emission maximum between ~ 320–360 nm, dependent on the protein environment and 2. flavins, which have two absorption maxima at ~ 360–390 nm and at ~ 440–470 nm (**Fig. 1**). The exact wavelengths of the absorbance maxima observed for protein-bound flavins depends on the nature of the flavin-binding site. The two most common biological forms of flavin are flavin mononucleotide (FMN, riboflavin 5' phosphate) and flavin adenine dinucleotide (FAD) (*see* Chapter 12 for structures). In the visible region, free FMN absorbs maximally at 373 nm and 445 nm (ε = 10,400 and 12,500 M^{-1} cm^{-1}, respectively) whereas free FAD has maxima at 375 nm and 450 nm (ε = 9300 and 11300 M^{-1} cm^{-1}, respectively) *(4,5)*. The fluorescence maxima for both

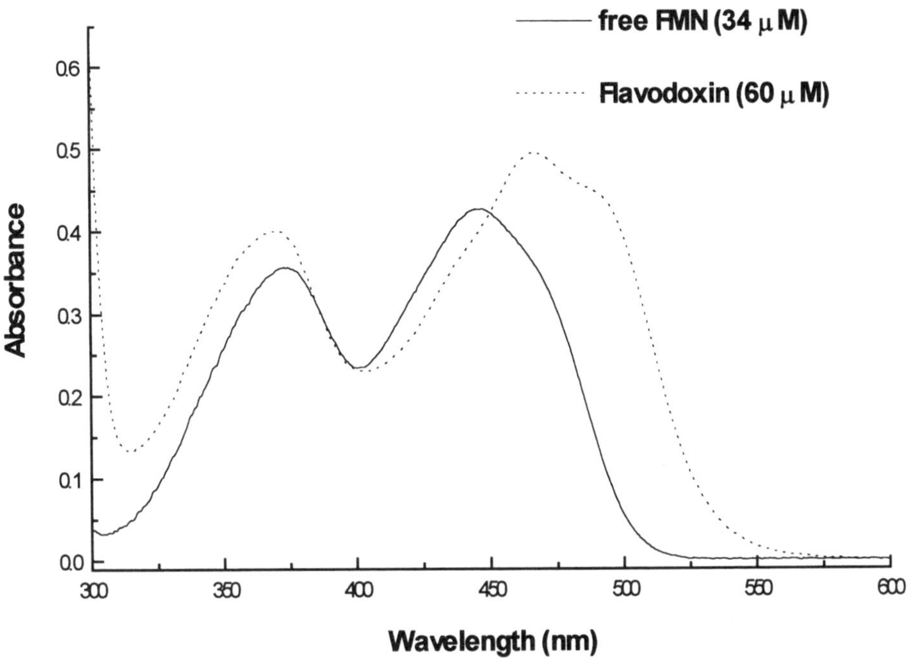

Fig. 1. Absorption spectra of flavin mononucleotide (FMN, 34 μ*M*—solid line) and the *E. coli* FMN-containing protein flavodoxin (60 μ*M*—dotted line). The environment of the FMN in the flavodoxin significantly perturbs its spectrum. Spectra were recorded on a Shimadzu UV 2101 Spectrophotometer. Solutions were made up in 100 m*M* potassium phosphate (pH 7.0).

forms are at 525 nm, but the Q_f is considerably greater for FMN (0.27) than for FAD (0.032) ***(6)*** due to quenching from the adenine ring in FAD. Fluorescence emission from irradiated free or protein-bound flavins occurs at longer (lower energy) wavelength. For flavoproteins, the emission maxima and Q_f are dependent on the nature of the flavin-binding site (**Fig. 1, Table 1**). For instance, the FMN in the flavodehydrogenase domain of the *Saccharomyces cerevisiae* L-lactate dehydrogenase flavocytochrome b_2 fluoresces maximally at 525 nm, and the fluorescence is very strongly quenched in the protein ($Q_f < 0.1$). The fluorescence from the noncovalently bound FMN in flavodoxins (small electron transferring proteins from various bacteria and lower eukaryotes) are extremely well quenched by the protein environment (Q_f values from ~ 0.003–0.013) (*see* **Subheading 3.2.1.**) ***(7)***. The FAD in eukaryotic lipoic acid dehydrogenase (E3 of the pyruvate dehydrogenase complex) has a fluorescence maximum at 520 nm and a Q_f of 0.11, whereas the FAD in L-amino acid oxidase is quenched to such a degree that there is essentially no fluorescence ***(1)***.

Table 1
Absorption and Fluorescence Properties of Flavins and Flavin-Containing Proteins in Oxidized and Reduced (red) States

Enzyme	Absorption maximum (nm)	Absorption coefficient (ε, M^{-1} cm^{-1})	Fluorescence emission maximum (nm)	Quantum yield (Q_f)
FMN	445	12, 500	525	0.27
FAD	450	11, 300	525	0.03
OYE (FMN)	462	10, 600	530	0.01
E3 (FAD)	455	11, 300	520	0.12
LO (FMN)	460	11, 300	—	—
LO (red)	360	4, 800	507	0.02
D-AAA (FAD)	455	11, 300	530	0.04
D-AAA (red)	365	5, 800	520	0.01

OYE = old yellow enzyme *(Saccharomyces carlsbergensis)*, E3 = lipoic acid dehydrogenase (porcine), LO = lactate oxidase *(Mycobacterium smegmatis)* and D-AAA = D—amino acid oxidase (porcine). OYE and LO contain FMN, E3 and D-AAA contain FAD. LO and D-AAA are examples of flavoproteins that are fluorescent in their reduced states. LO is nonfluorescent when oxidized ***(1,7,31)***.

The characteristics of the fluorescence from protein-bound flavins can be used to investigate a number of phenomena. For instance, it is possible to distinguish between different flavin forms, to quantify the flavin, to determine the integrity of the flavin binding site and examine the extent to which the surrounding protein structure quenches the flavin fluorescence, to monitor reaction kinetics, and to investigate molecular interactions. Examples of useful techniques in the analysis of flavoproteins are given in **Subheading 3.**

2. Materials

2.1. Spectrometer

An essential requirement is obviously the fluorescence spectrometer (fluorimeter). These are widely available from a number of manufacturers, including Shimadzu (Tokyo, Japan), Kontron, Hitachi (Irvine, CA), Perkin-Elmer (Norwalk, CT) and Jasco (Eason, MD). The levels of sophistication of these instruments vary widely, along with their price. Many standard instruments can operate in "kinetic" mode—allowing the collection of fluorescence data over time (seconds to minutes) to monitor processes such as enzyme-dependent conversions of fluorescent substrates. The fluorescence mode is also available on various stopped-flow spectrometers, including models from Applied Photophysics (Leatherhead, UK) and Hi-Tech (South Yorkshire, UK). This facility allows the monitoring of fluorescence changes over millisecond time scales and can be useful for following the changes in fluorescence associated with e.g., flavin binding during protein refolding *(8)*.

Fluorimeters are generally of similar design to absorbance spectrophotometers. However, there are some important differences. Since fluorescence is proportional to the intensity of the incident light, the light source is usually a more intense one than used for absorption spectrophotometry, i.e., a mercury or xenon arc, as opposed to the tungsten or deuterium lamps used in absorption instruments. Mercury arcs provide good excitation, but the fact that the number of photons emitted varies with excitation wavelength can be disadvantageous in the recording of excitation spectra (*see* **Subheading 3.1.1.**). Xenon arcs are much brighter and are widely used in scanning fluorimeters *(3)*.

2.2. Cuvet

The sample should be contained within a cuvet, commonly of 1 cm pathlength and of volume 1–3 mL (although smaller and larger cuvets are commercially available). Fluorescence cuvets should be made of silica (quartz) and are generally polished on all four faces. This may mean that they cost rather more than cuvets suitable for absorption spectrophotometry. The fluorescence cuvet should be cleaned routinely with water and then acetone or ethanol, and polished with a clean lens tissue. The cuvet sits in a sample compartment that usually has an open entry port facing the excitation light beam and an exit port for the emission that is at right angles to the excitation beam.

3. Method

3.1. Collecting Fluorescence Spectra of Flavoproteins

3.1.1. Excitation Spectra

The fluorescence excitation spectra of flavoproteins should be virtually identical to their absorption spectra (**Figs. 1** and **2**). However, non-corrected excitation spectra may be rather different from absorbance spectra (*see* **Note 6**). An excitation spectrum should be run for comparison with the absorption spectrum and in order to select the best excitation wavelength(s) for collection of emission spectra. As with emission spectra, the flavoprotein sample should be of the highest possible purity and diluted into distilled water or a buffer solution of desired pH (e.g., Tris.HCl or sodium phosphate). Particulate matter should be removed by brief centrifugation. Passing the sample/buffer through a Millipore filter (0.45 μ*M*) using a syringe may also remove dust and improve the quality of the data collected. The cuvet should be cleaned as described in **Subheading 2.2.** The quantity of flavoprotein required for an excitation

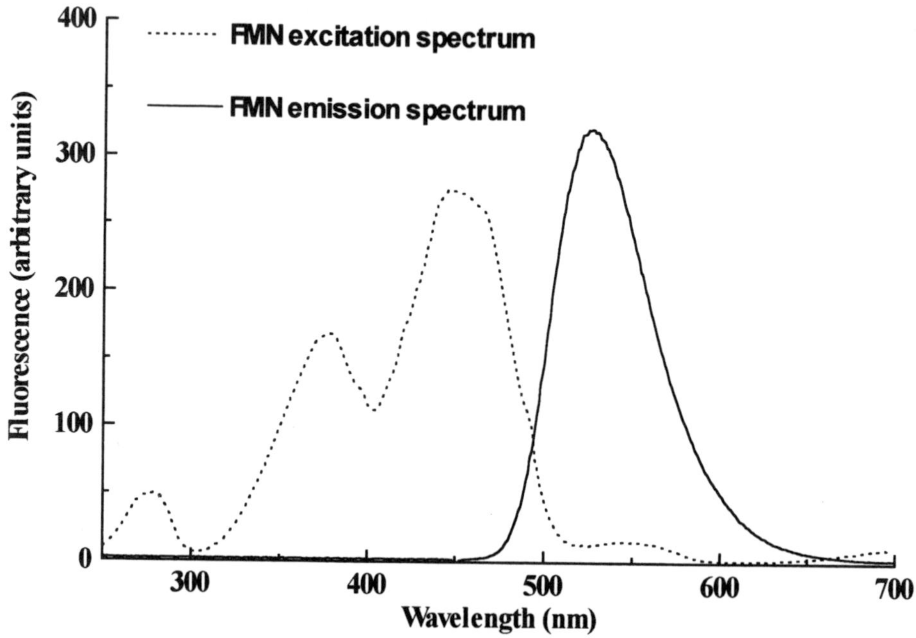

Fig. 2. Fluorescence excitation (dotted line) and emission (solid line) spectra from FMN (4 μ*M* in 100 m*M* potassium phosphate, pH 7.0). For the excitation spectrum, emission was measured at 525 nm with spectral band widths at 3 nm (excitation) and 1.5 nm (emission). For the emission spectrum, excitation was at 450 nm with spectral band widths at 1.5 nm (excitation) and 3 nm (emission). Spectra were recorded on a Shimadzu RF 5301 Spectrofluorimeter. The excitation spectrum was not corrected (*see* **Note 6**) and excitation at 450 nm overlaps with the emission spectrum, emphasizing the benefit of using an exciting wavelength lower than the λ_{max} (*see* **Subheading 3.1.2**).

spectrum should be relatively small (around 1 μ*M*)—and the operator should investigate 3 or 4 flavoprotein concentrations in this range. There is no relationship equivalent to the Beer-Lambert law for fluorescence measurements, and the results obtained are dependent on a number of factors—including the instrument used. The spectral band widths used for excitation and emission beams also have dramatic effects on the intensity of signals collected (*see* **Note 2**). The spectral band width is a property of the fluorimeter used, and is defined as the band of wavelengths containing the central half of the entire band of wavelengths passed by the exit slit of the monochromator ***(9)***. The size of the excitation and emission slits on the instrument can be varied and may be calibrated as "spectral band width" (in nm) or simply as "slit width" (in mm)—where the latter is a physical measurement of the width of the slit. The relationship beween these two values should be available in the instrument manual, but a slit width of 1 mm usually corresponds to a spectral band width of 2–4 nm. For flavoprotein excitation spectra the operator should keep the excitation spectral band width minimal (e.g., ≤ 3 nm) and have a wider emission band width (e.g., ≥ 5 nm). The spectrum of the protein-free buffer should be read and this should be subtracted from that of the protein solution to obtain a "true" excitation spectrum.

3.1.2. Emission Spectra

Having collected an excitation spectrum of the oxidized flavoprotein, the optimal excitation wavelength (i.e., that showing maximal excitation/absorbance) required to obtain the maximal fluorescence may be selected. In the case of the *E. coli* FMN-containing flavodoxin this would be at 466 nm (**Fig. 1**), but may be considerably different for other flavoproteins (**Table 1**). However, it is found for many fluorophores that the excitation and emission spectra overlap significantly. This may well be the case with the flavoprotein of interest (since flavin absorbance and emission spectra are broad) and could mean that if the excitation maximum is used, light scattering leads to distortion of part of the emission spectrum. In order to record a complete, undistorted emission spectrum, the best approach may be to select a lower wavelength for excitation (e.g., ~ 430 nm for most flavoproteins). Although this will result in less intense emission, it should not affect the form of the spectrum *(1)*.

The same precautions regarding cuvets, buffer and flavoprotein should be taken as described for the excitation spectra. In contrast to conditions for excitation spectra, the operator should select a relatively broad excitation spectral band width for emission spectra (to maximize excitation) (e.g., ≥ 5 nm) and a narrower emission spectral band width (e.g., ≤ 3 nm). An appropriate quantity of flavoprotein required may be between ~ 0.5–2 μ*M*. The extent to which the flavin fluorescence is quenched by the protein will affect the concentration of protein required. If sample is limited, then any problems with low signal intensity can be addressed by widening the excitation/emission spectral band widths. However, the operator should be aware that increasing the spectral band widths too much can not only result in the collection of non-flavin-specific fluorescence, but may also affect the sensitivity of the emission measurement (*see* **Notes 1** and **2**). As with excitation, a blank (i.e., buffer only) emission spectrum should also be run and this should be subtracted from the sample spectra.

Fluorescence emission can also prove a useful tool during the purification of flavoproteins from various sources—particularly if the protein of interest is not expressed at high levels and is not easily detectable by a visible yellow color in fractions from column chromatography steps. Samples can be read directly (with excitation at ~ 450 nm and emission at ~ 540 nm) to detect those fractions containing flavin. If this technique can be coupled with another specific assay for the protein in the fluorescent fractions, then it can expedite the purification process.

3.2. Flavoprotein Denaturation and Renaturation

As indicated in **Subheading 1.2.**, the fluorescence from protein-bound flavins may be severely quenched by energy transfer to the surrounding protein matrix. For instance, in many flavoenzymes, aromatic amino acids (e.g., tyrosine, tryptophan) help to stabilize the binding of the flavin by providing aromatic stacking interactions with the isoalloxazine ring system (e.g., Trp677 with the FAD of mammalian cytochrome P450 reductase ***(10)*** and Tyr308 with the FAD of pea ferredoxin $NADP^+$ reductase ***(11)***). The fluorescence quenching of bound flavins can be of great benefit in the study of flavoprotein denaturation/renaturation. During the unfolding of flavoproteins, the secondary and tertiary structure of the protein is disrupted and interactions with flavin break down, usually leading to large increases in the flavin fluorescence. For non-covalently bound flavoproteins, this usually reflects flavin dissociation. For covalently bound flavoproteins, changes in fluorescence properties may also be observed—but these are likely to reflect increased solvent accessibility/loss of protein-flavin contacts as opposed to complete removal of the flavin from the polypeptide chain. Protein unfolding/refolding can be monitored by both near UV (tertiary structure) and far UV (secondary structure) circular dichroism (*see* Chapter 8) and by changes in the fluorescence of aromatic amino acids. Alteration in flavin fluorescence provides a method of assessing the integrity of the flavin-binding site and a useful means of comprehending the processes (and their velocity) that occur during the folding and unfolding of a flavoprotein. Flavin-containing active sites may be very sensitive to disruption ***(12,13)***, although certain flavoproteins (e.g., the FMN-containing flavodoxins) bind their flavins very tightly (*see* **Subheading 3.3.1.**).

3.2.1. Flavin Fluorescence During Flavoprotein Denaturation

A variety of methods are available to disrupt protein structure. The most popular techniques involve chemical agents (usually guanidine hydrochloride [GdnHCl] or urea) or thermal treatment. However, temperature changes affect fluorescence intensity and over-heating often leads to irreversible protein aggregation. Therefore chemical treatment may be a wiser choice. Urea or GdnHCl should be the highest grade available (Gibco-BRL [Paisley, Scotland], Sigma [St. Louis, MO], Calbiochem Novabiochem [San Diego, CA] and other suppliers offer high quality denaturants) to avoid the presence of contaminants

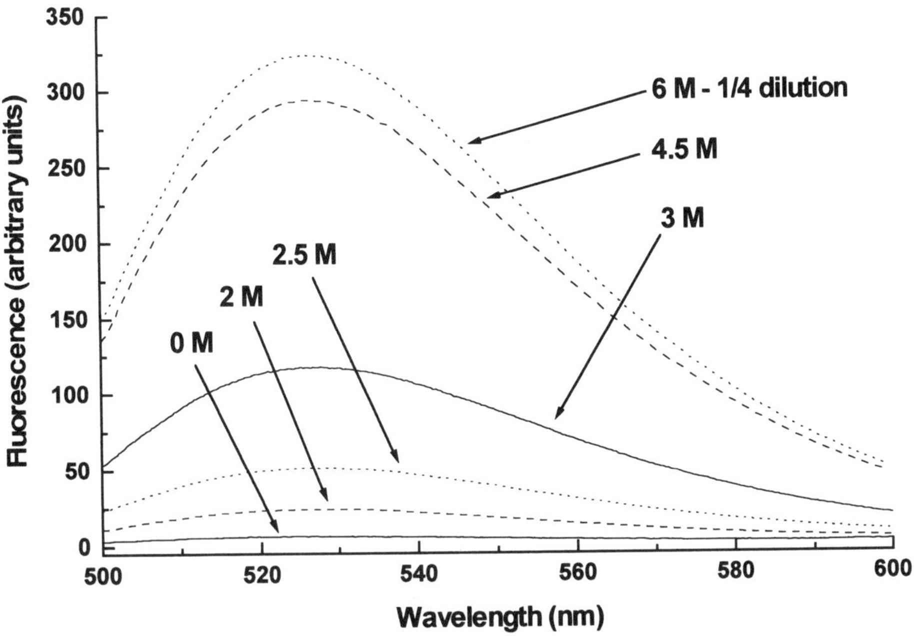

Fig. 3. Fluorescence emission spectra for the *E. coli* FAD-containing flavodoxin reductase (5 μ*M*) at different concentrations of guanidinium chloride (GdnHCl). Spectra were recorded in 10 m*M* potassium phosphate (pH 7.0) at 0 *M* (full line), 2 *M* (dashed line), 2.5 *M* (dotted line), 3 *M* (full line), 4.5 *M* (dashed line) and 6 *M* (dotted line) GdnHCl. The GdnHCl induces unfolding of the enzyme and releases quenching of the FAD fluorophore by the enzyme. The fluorescence is enhanced so much at 6 *M* GdnHCl that the enzyme had to be diluted four-fold to keep the spectrum on scale. Excitation was at 450 nm with spectral band widths at 3 nm (excitation) and 5 nm (emission). Spectra were recorded on a Shimadzu RF 5301 Spectrofluorimeter.

that may act as fluorescence quenching agents. Solutions of the compounds should be made up freshly (breakdown products, particularly cyanate from urea, can affect the data) and in the same buffer used for the enzyme. The concentration of flavoprotein used for the emission spectra and the excitation wavelength should be determined as described in **Subheading 3.1.1.** However, the operator should be aware that there may be a considerable (perhaps > 10-fold) increase in flavin fluorescence as the protein unfolds (**Fig. 3**) and that the use of a narrower emission and/or excitation slit may be required to keep all data on scale without recourse to sample dilution. There may also be large changes in the λ_{max} of fluorescence between free- and protein-bound flavin (e.g., ***14***). Samples of the flavoprotein should be made up to the same concentration in the desired buffer and in buffered solutions of the denaturant ranging from 0–6 *M* (for GdnHCl—from an 8 *M* stock) or 0–8 *M* (for urea—from a 10 *M* stock). It is prudent to incubate all samples for the same time period to allow equilibration prior to taking readings (e.g., 15 min at room temperature), and so it may be beneficial to "stagger" the dilutions of enzyme stocks into preprepared solutions of denaturant—e.g., preparing one sample every 10 min. At the end of the incubation period, samples are transferred to the fluorimeter and flavin fluorescence recorded using the excitation wavelength determined (**Subheading 3.1.1.**) and either measuring emission at a predetermined λ_{max}, or scanning fluorescence emission over a wavelength range. The latter approach is advisable, since the λ_{max} for emission and pattern of the emission spectrum may change somewhat as the flavin environment is altered during protein unfolding. With excitation at 430 nm, an emission range of 500–600 nm should encompass > 90% of flavin fluorescence emission.

Having prepared solutions and measured the flavin fluorescence, it may be useful to measure aromatic amino acid (chiefly tryptophan) fluorescence using the same sample, in order to compare the alteration of protein tertiary structure with the changes in flavin properties. Appropriate excitation wavelength is ~ 290 nm and the majority of the emission will fall within the range 320–400 nm. As with flavin fluorescence, there may be considerable increase in aromatic fluorescence during unfolding. Measurement of aromatic fluorescence from the same sample may require alteration of spectral band widths to keep signal intensity on the same scale. A direct plot of the emission intensity at a selected wavelength against concentration of denaturant may allow direct comparison of the stability of the overall tertiary structure of the protein with that of its flavin-binding site(s) ***(12)***. More detailed texts deal with the determination of the energetics of the unfolding process (e.g., ***15***).

3.2.2. Flavin Fluorescence During Flavoprotein Renaturation

The renaturation of completely unfolded proteins and rebinding of dissociated flavins may be difficult (even impossible) to achieve and very complex to analyze. Chaperone proteins may be required and the kinetics of refolding may be difficult to analyze due to heterogeneity of the unfolded state ***(16)***. However, at its simplest, fluorescence measurement of flavin reassociation can be made after direct dilution of denaturant-treated samples (prepared as described in **Subheading 3.2.1.**) into denaturant-free buffer. Ideally, if the unfolding of the flavoprotein is freely reversible, this will lead to the regeneration of the fluorescence properties of the native form. The process can also be studied on a stopped-flow time-scale. For instance, the Applied Photophysics instrument allows the mixing of samples in a 10:1 ratio, and hence the dilution of denaturant from e.g., 2–0.18 *M* in the mixing chamber. However, the processes are likely to be rather more difficult to optimize and approaches such as the following may be beneficial for successful reactivation:

1. Using stepwise dilutions of unfolded flavoprotein into buffered solutions of decreasing [denaturant] or dialysis of the unfolded flavoprotein into the refolding buffer.
2. Addition of an excess of the appropriate free flavin to the refolding solution(s) (perhaps ~ 100 μ*M*). These should be removed by gel filtration prior to fluorescence measurement.
3. Repeat measurements using a range of different concentrations of the flavoprotein during refolding.
4. Addition of enzyme systems that may promote refolding (e.g., GroEL/GroES/ATP or protein disulfide isomerase ***(17)***).

It should be remembered that the flavin (and tryptophan) fluorescence will probably be markedly reduced on dilution/refolding, and that wider spectral band widths for excitation and/or emission may be needed. Also, rapid events in protein refolding may be followed by much slower changes that may affect the fluorescence over several minutes. It is therefore advisable to monitor fluorescence changes over this time period.

3.3. Flavin Quantitation and Identification

The well-characterized fluorescence properties of FAD and FMN allow them to be easily quantified, and methods exist by which the relative quantities of these two most common flavins can be determined in mixtures. This has particular importance for such enzymes as cytochrome P450 reductase and nitric oxide synthase, which contain both FAD and FMN ***(18,19)***. Two useful methods for FAD/FMN quantitation and discrimination are given in the following section.

3.3.1. Flavin Determination by Fluorescence Quenching with Apoflavodoxin

Flavodoxins are a group of small, structurally well-characterized FMN-containing proteins that act as electron shuttles. They are widespread in bacteria and have also been isolated from lower eukaryotes and algae ***(8)*** This method for flavin quantitation relies on the fact that the apoenzymes (flavin-free forms) of flavodoxins from the anaerobic bacteria *Megasphaera elsdenii* (ATCC 25940) and *Clostridium pasteurianum* (ATCC 6013) bind FMN rapidly and very tightly, but riboflavin and FAD extremely weakly ***(20)***. FMN binding to the apoflavodoxin quenches its fluorescence strongly (**Fig. 4**). Not all flavodoxins are as selective as those above and it is therefore best to purify these flavodoxins directly from *M. elsdenii* or *C. pasteurianum* ***(21,22)***. This is a relatively simple process, since the flavodoxins are colored yellow and are acidic proteins that bind well to anion exchange resins ***(8)***.

1. FMN is removed from the purified flavodoxin by mixing a solution of the protein (5 mg/mL—determined using ε_{445} = 10.2 mM^{-1} for *M. elsdenii* or

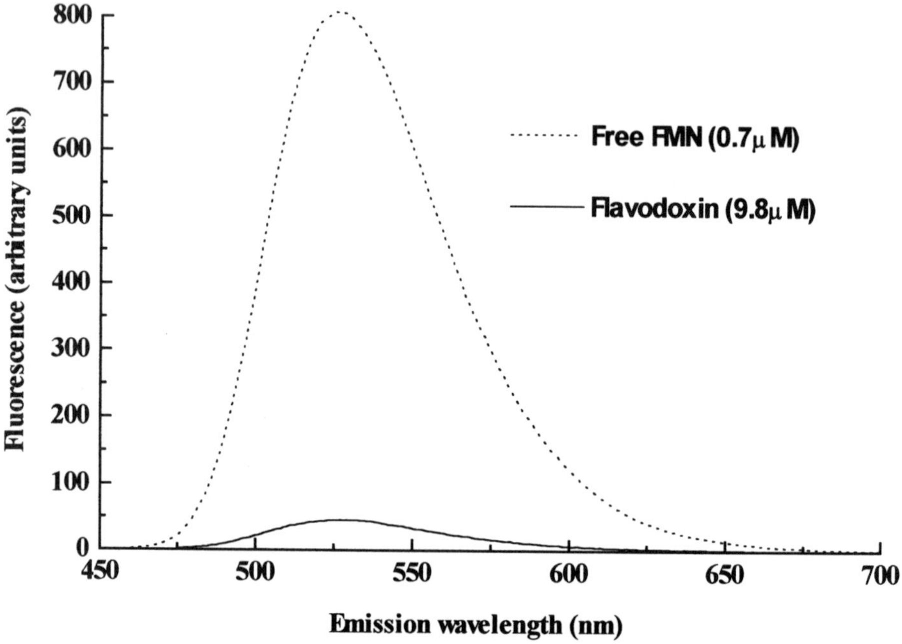

Fig. 4. Fluorescence emission spectra from flavin mononucleotide (FMN, 0.7 m*M*—dotted line) and the *E. coli* FMN-containing protein flavodoxin (9.8 μ*M*—solid line). The FMN is very strongly quenched in the flavodoxin and its fluorescence emission intensity is much reduced. Spectra were recorded on a Shimadzu RF 5301 Spectrofluorimeter with excitation at 450 nm. Spectral band widths were 3 nm (excitation) and 5 nm (emission), respectively. Solutions were made up in 100 m*M* potassium phosphate (pH 7.0).

ε_{443} = 10.4 mM^{-1} for *C. pasteurianum* flavodoxin) in cold protein buffer (10 m*M* potassium phosphate buffer, 0.3 m*M* EDTA [pH 7.0]) with cold trichloroacetic acid (50% w/v) to give a final acid concentration of 5% (w/v).

2. The solution is stored in darkness at 4°C for 5 min and is then centrifuged at 10,000*g* for 10 min.
3. The supernatant contains free FMN, which is removed, and the apoprotein pellet is washed free of residual FMN by resuspending in 5% (w/v) trichloroacetic acid plus 0.3 m*M* EDTA (0.4 mL/mg apoprotein).
4. The apoprotein is centrifuged as before and dissolved in a minimal volume of protein buffer, and dialyzed overnight at 4°C against the same buffer.
5. The apoprotein is stored at 4°C until use.
6. The concentration of the apoenzyme is determined by titration with FMN. High quality FAD is available, but commercial preparations of FMN contain impurities that are not bound by the apoflavodoxins. It is advisable to purify the FMN by chromatography on DEAE cellulose. An alternative (superior) method is to purify commercial FAD on DEAE cellulose and then to hydrolyze it to FMN

with *Naja naja* snake venom (Sigma) which contains a phosphodiesterase, as outlined in 14 below.

7. The solution of apoflavodoxin is diluted to approximately 20 μ*M*, using the coefficients ε_{278} = 26.7 mM^{-1} and ε_{282} = 25.2 mM^{-1} for the *M. elsdenii* and *C. pasteurianum* apoflavodoxins, respectively ***(20)***. These coefficients give the absolute quantity of apoflavodoxin, but that proportion that binds FMN must be determined by titration with FMN.
8. FMN titration is performed by successive additions of 10–15 μL aliquots of the approx. 20 μ*M* apoflavodoxin stock to a 2 mL solution of 1 μ*M* FMN made up in 10 m*M* sodium acetate buffer (pH 6.0) containing 0.2 *M* NaCl, within a 1 cm path-length fluorescence cuvet.
9. After each addition, the flavin fluorescence is recorded with excitation at 445 nm and emission at 525 nm.
10. The titration is continued until further additions of apoflavodoxin do not cause any further decreases in flavin fluorescence.
11. These data are then corrected for dilution and the fluorescence intensity plotted against the volume of apoflavodoxin added. The "breakpoint" in the curve where fluorescence is maximal is extrapolated to the *x*-axis to determine the volume of flavodoxin that is equivalent to 1 μ*M* FMN. This allows the calculation of the "active" apoflavodoxin concentration.
12. The standardized apoflavodoxin stock is then used in titrations with samples of unknown flavin content and composition. As with the flavodoxins, flavins can be prepared from sample enzymes or other biological samples by trichloroacetic acid extraction.
13. The acetate buffer system is again used and the sample is either diluted to approximately 1 μ*M* flavin, or (if it contains much less than this to start with) the apoflavodoxin solution is diluted to such an extent that a number of data points can be obtained for that part of the curve where there is an essentially linear decrease in flavin fluorescence with increase in apoflavodoxin concentration. For samples of low flavin content, it may be necessary to increase excitation and/or emission spectral band widths.
14. In mixtures containing both FAD and FMN, a second titration is performed after addition of the *Naja naja* venom to an identical test solution.
15. The venom is made up to a concentration of 10 mg/mL in 10 m*M* potassium phosphate buffer (pH 7.0) and 10 μL of the venom is added to the test solution.
16. The *Naja naja* phosphodiesterase hydrolyses FAD to FMN, and this results in a considerable increase in the fluorescence of the sample (*see* **Subheading 1.2.**).
17. After the fluorescence increase is seen to be complete (~ 2 min), the apoflavodoxin titration is repeated. The quantity of FAD is determined from the difference in the end-point of the titrations with and without the phosphodiesterase.

This method is useful for solutions containing (≥100 n*M* flavin, providing that the [FMN] in the solution is > 20 n*M*. It can be used successfully for crude flavoprotein-containing preparations ***(23)***.

3.3.2. Flavin Determination by pH-Dependent Fluorescence

An alternative fluorescence-based method for FAD/FMN quantitation exploits the distinct pH-dependent fluorescence properties of these flavins *(22)*.

1. Stock solutions of pure FAD and FMN (*see* **Subheading 3.3.1.** for preparation) at 10 μ*M* are made up fresh in assay buffer (100 m*M* potassium phosphate, pH 7.7). All solutions are protected from light by covering tubes with aluminum foil.
2. Samples of the flavoproteins of interest are diluted with assay buffer to final concentrations of between 10–50 n*M*. The concentration can be estimated with sufficient accuracy by measurement of absorbance of the flavoprotein stock and assuming $\varepsilon_{450} = 11\ mM^{-1}\ cm^{-1}$.
3. Aliquots (3 mL) of the flavoprotein sample are placed into sealed glass tubes, covered with aluminum foil and transferred to a bath of boiling water for 3 min.
4. The samples are then centrifuged at 10,000 *g* for 15 min at 4°C to precipitate denatured protein.
5. 2 mL of the supernatant are transferred to a 1 cm path-length fluorescence cuvet.
6. The fluorescence is measured with excitation set at 450 nm and emission at 535 nm, ensuring that the sample has cooled to ambient temperature (approx 20°C).
7. 0.2 mL of 1.0 *M* HCl is added to the cuvet to reduce the pH to 2.6, and the fluorescence is remeasured.
8. Fluorescence readings (under the same conditions) are made at pH 7.7 and 2.6 for samples of pure FAD and FMN. Readings should be taken for three different concentrations (20, 40 and 60 nM) of both FAD and FMN. A constant for the fluorescence of FMN at pH 7.7 can then be calculated from this data using: $F_{7.7} = N_{7.7}$ [FMN], where $F_{7.7}$ is the fluorescence value measured for pure FMN at pH 7.7 and $N_{7.7}$ is the "fluorescence constant" (in units of $M^{-1}\ cm^{-1}$) for FMN at this pH.
9. The process is repeated to determine fluorescence constants for FMN at pH 2.6 ($N_{2.6}$) and for FAD at pH 7.7 ($D_{7.7}$), and pH 2.6 ($D_{2.6}$). Determinations of standard flavin solutions at three different concentrations maximizes the accuracy of the procedure and enables determination of standard errors.
10. For the sample(s) of interest, the fluorescence values at pH 7.7 and 2.6 are given by: $F_{7.7} = N_{7.7}$ [FMN] + $D_{7.7}$ [FAD] and by $F_{2.6} = N_{2.6}$ [FMN] + $D_{2.6}$ [FAD]. The values of the fluorescence constants for FMN/FAD are now known and hence the concentrations of FMN and FAD in the sample(s) can be determined.

The method is useful for dilute flavin-containing solutions (100 n*M* or less), where absorbance is negligible and fluorescence is directly proportional to the flavin concentration. It is a rapid method that can be useful for quantitating and identifying the nature of the cofactor in a single flavin-containing protein (e.g., ***24***). It is more accurate for determinations with relatively pure flavoprotein samples. Due to instrument variation, the fluorescence constants are dependent on the fluorimeter used.

3.4. Other Techniques

3.4.1 Flavin Fluorescence Polarization

Fluorescence polarization (or anisotropy) is a potentially very powerful tool for the study of molecular interactions *(3)*. For flavoproteins, it can be particularly useful in the study of the binding of the flavin fluorophores to the proteins. The principle of the technique is that a solution of the flavin is excited with plane polarized light and that molecules with absorption moments parallel to the plane of polarization are excited preferentially. The emission from the excited flavin will also be polarized, providing that the flavin remains stationary during the lifetime of its excited state. If the excited flavin rotates or tumbles, then the polarization of emitted light is decreased. The fluorescence polarization value of a molecule relates to its rotational relaxation time, which is the time taken for the molecule to rotate through 68.5°. Under identical reaction conditions of viscosity and temperature, this is directly related to the "molecular volume" of the fluorophore—a value that can be altered for a flavin by such phenomena as protein association or dissociation, and flavoprotein conformational change.

Successful use of the technique requires that the fluorimeter is fitted with a filtering apparatus to polarize the excitation and emission beam. Many manufacturers (e.g., Kontron) can provide such accessories and specialized instruments for measuring fluorescence polarization are also available (e.g., the Beacon system™ from the PanVera Corp., Madison, WI). Buffers and other reagents used should also be of the highest quality to ensure minimal background fluorescence.

3.4.2. Flavin Fluorescence Dynamics

The rates at which overall flavin fluorescence and polarization of flavin fluorescence decay can be monitored to investigate such phenomena as polarity of the flavin binding site (e.g., ***25***), rates of flavin binding and affinities of proteins for flavins. The dynamics of fluorescence decay are measured over very short timescales (picoseconds to nanoseconds), which are beyond the limitations of conventional stopped-flow apparatus, and require the use of laser excitation. A detailed account of the methods and applications of fluorescence dynamics is beyond the scope of this chapter. However, there are numerous accounts of the use of "flavin dynamics" in the study of flavoproteins, and these provide descriptions of the machinery required and references to the theory involved (e.g., ***25–28***)

3.4.3. Quenching of Flavin Fluorescence

The accessibility of flavins within a protein and how this accessibility is altered by external factors can be investigated by the extent to which certain chemicals (quenchers) are able to quench the fluorescence from flavins. Commonly used quenchers include compounds such as potassium iodide, caesium chloride, and acrylamide *(29)*. The iodide and chloride are ionic quenchers and cannot penetrate into the hydrophobic core of a protein, whereas acrylamide (which is a very efficient quencher of tryptophan fluorescence) is hydrophobic and nonpolar, and can penetrate protein regions inaccessible to iodide or chloride. The effects of quenchers can be determined by titration of the quenchers into solutions of the flavoprotein of interest and recording the flavin fluorescence intensity at each point in the titration, as described in **Subheading 3.1.2.** It may also be of interest to determine simultaneously the quenching effect on protein tryptophan residues (*see* **Subheading 3.2.1.**). High concentrations of the quenchers may be required to achieve good quenching of fluorescence (up to, or more than 1 *M*). If iodide is used, the buffer should also contain 10 m*M* sodium thiosulfate to prevent the formation of reactive I_3^- (which also absorbs in the near-UV region) *(29)*.

The quenching data can be analyzed using the Stern-Volmer equation:

$$F_0/F = 1 + K_{SV}\,[Q]$$

where F_0 and F are the fluorescence values recorded in the absence and presence of quencher, respectively, K_{SV} is the Stern-Volmer constant and $[Q]$ is the concentration (*M*) of the quencher. If all fluorophores in a protein are equally accessible to the quencher, then a plot of F_0/F vs $[Q]$ should be linear. However, curvature may be observed at higher $[Q]$ if e.g., there are non-equivalent flavins in a protein, as seen in P450 reductase or nitric oxide synthase *(18,19)*. The greater the value of the Stern-Volmer constant, the greater is the accessibility of the fluorophore to the quencher. The fraction of the fluorophore accessible to a quencher can be calculated from a modified form of the Stern-Volmer plot (the Lehrer plot) according to the equation:

$$F_0/\Delta F = 1/K_Q.f_a.[Q] + 1/f_a$$

where ΔF is the change in fluorescence intensity through quenching, KQ is the modified quenching constant and f_a is the fraction of the fluorophore accessible to the quencher *(30)*. If F_0/DF is plotted vs $1/[Q]$, then the gradient of the plot is $1/KQ.f_a$ and the y-axis intercept is $1/f_a$.

Fluorescence quenching is a useful technique for investigation of e.g., structural comparisons between related flavoenzymes, and for probing the relative accessibilities of flavins in substrate- or inhibitor-bound enzymes compared with the ligand-free forms.

3.4.4. Fluorescence from Reduced Flavins

Most studies of flavoproteins report only the fluorescence properties of the oxidized forms of the cofactors. Indeed, it is often assumed that reduced flavoproteins are non-fluorescent. However, numerous reduced flavoproteins do exhibit fluorescence ***(30)*** Flavins in the partially (semiquinone) and fully reduced (hydroquinone) state can fluoresce, and this may be of interest for investigating the flavin exposure or measuring electron transfer rates through catalytically important states of the enzyme. Indeed, certain reduced flavoenzymes (such as lactate oxidase) may fluoresce in the reduced state, whereas the oxidized flavin is totally quenched by the protein environment ***(31)***. The absorbance properties of the reduced flavoproteins will be considerably different from those of the oxidized form, and an excitation spectrum of the reduced enzyme of interest should be taken to determine the optimal excitation wavelength. Rapid reoxidation of the flavoprotein may prove a problem and in such cases the buffer should be deoxygenated and the enzyme reduced anaerobically.

3.5. Examples of the Study of Flavoproteins by Fluorescence

There are an enormous number of examples of the use of fluorescence spectroscopy in the analysis of flavoproteins to be found in the scientific literature. Following are a small number of examples of fluorescence studies reporting different properties of flavoproteins.

Fluorescence spectra have been reported for a large number of flavoproteins, including *Escherichia coli* thioredoxin reductase ***(32)***, the *E. coli* glutathione reductase (GOR) ***(33)*** and the particularly complex emission spectrum from the *Pichia pastoris* methanol oxidase ***(34)***. The fluorescence from bound flavins is very strongly quenched in many flavoproteins. However, mutations can have significant effects on the fluorescence, such as the large FAD fluorescence increases observed (i.e., decreased quenching) in Y177 mutants of *E. coli* GOR ***(33)***. The fluorescence enhancements in mutants Y177F/G of GOR is also associated with conformational changes that result in increased affinity (lower K_m) for reduced glutathione ***(35)***. Excitation spectra have been reported from e.g., mercuric reductase (FAD) from *Pseudomonas aeruginosa* ***(36)*** and the *Pichia pastoris* methanol oxidase ***(34)***. Studies on the fluorescence emission spectra of reduced flavoproteins are relatively rare. However, enzymes such as lactate oxidase, D-amino acid oxidase, oxynitrilase and flavodoxins (various forms) have been shown to fluoresce in sodium-dithionite-, substrate- or photo-reduced forms ***(30)***. Flavin dynamics have also been investigated for the reduced forms of lactate oxidase and *M. elsdenii* flavodoxin ***(26)***.

Fluorescence-based methods are used frequently for the quantitation and identification of flavins. For instance, the apoflavodoxin method ***(20)*** has been used to determine the presence and amount of FMN in rabbit cytochrome P450 reductase ***(37)*** and to examine the flavin content of the pyruvate dehydrogenase complexes from *Azotobacter vinelandii* and *E. coli* ***(38)***. The pH-dependent method ***(23)*** has been used to examine the specificity of a technique for the removal of one of the two non-equivalent flavins of xanthine dehydrogenase *(39)*. It has also been employed in study of the flavin fluorescence increases during urea- and GdnHCl-mediated unfolding of mammalian P450 reductase *(40)* and the fatty acid oxygenase flavocytochrome P450 BM3 from *Bacillus megaterium* *(12)*, demonstrating the weaker binding of FMN compared with FAD in these enzymes.

Time-resolved fluorescence and fluorescence polarization have been used to examine the properties of the FAD in human erythrocyte GOR. The short fluorescence polarization decay correlation time indicated there to be motion of a small FAD-binding segment of the enzyme, independent of the rest of the protein *(27)*. Analysis of the fluorescence lifetimes of protein-bound and free FMN for *Desulfovibrio vulgaris* flavodoxin *(41)* and *E. coli* chorismate synthase *(28)* allowed calculation of dissociation constants for oxidized FMN cofactor for both proteins. Further studies of the flavin fluorescence dynamics of a number of microbial flavodoxins indicated that reduced FMN is immobilized much more strongly in the protein matrix than the oxidized cofactor *(42)*.

Quenching of flavin fluorescence by iodide has been used to probe solvent accessibility of the FAD in wild-type and tail-deleted *A. vinelandii* lipoamide dehydrogenase. The FAD in the tail-deleted form was much more accessible to iodide, in agreement with a model in which the C-terminal portion of the enzyme shields the flavin from solvent *(43)*. Iodine quenching has also been used to investigate the relative accessibilities of FAD and FMN in mammalian P450 reductase. The results indicated that the FAD is more deeply embedded in the enzyme than is the FMN *(44)*. Domain rotation within *E. coli* thioredoxin reductase was investigated by analysis of the FAD fluorescence-quenching properties of the $NADP^+$ analogue aminopyridine adenine dinucleotide phosphate ($AADP^+$). Wild-type thioredoxin reductase showed a similar degree of fluorescence quenching by $AADP^+$, whether or not the enzyme was pre-treated with the sulfhydryl reagent phenylmercuric acetate (PMA). By contrast, PMA treatment of mutant C138S (which removes one of the two cysteines of the active-site disulfide) in order to modify the other exposed cysteine of the pair (C135) results in a 9.5-fold increase in FAD fluorescence, and this fluorescence is significantly quenched by subsequent addition of $AADP^+$ *(45)*. These data are consistent with the model in which the domains of thioredoxin reductase can rotate to place either the NADPH-binding site or the redox active disulfide adjacent to the flavin.

4. Notes

The operator should be aware of certain problems that can cause confusion and lead to irreproducibility of data. Careful experimentation will allow these problems to be overcome easily.

1. The "inner filter effect" occurs when the exciting light is strongly absorbed by the fluorophore solution or, less frequently, when the emitted light is absorbed. If the fluorophore concentration is too high, then most incident light may be absorbed near the front face of the cuvet and very little fluorescence emission (detected at right angles from the center of the sample) will be recorded *(3)*. Under such conditions, dilution of the sample may lead to increased fluorescence detection. A "front face" attachment may be available for the fluorimeter used, to allow measurement of fluorescence from highly absorbing solutions. However, this should not be necessary for flavoproteins.

   There is no direct equivalent to the Beer-Lambert law for fluorescence measurements, but in dilute solutions, fluorescence is proportional to
   a. the incident light intensity,
   b. the absorptivity of the fluorophore and
   c. the quantum yield (Q_f; *see* **Subheading 1.1.**).

   However, if the absorbance of the solution is > 0.05 at the excitation wavelength then relationship between fluorescence intensity and concentration of the fluorophore will be non-linear, and a calibration curve will be required. The operator should establish the linearity of response over a range of concentrations of the flavoprotein of interest. The flavin fluorescence response should be linear between ~ 5 n*M* and 5 µ*M*, and this is a sensible working range for flavoprotein analysis.
2. Due to the relatively broad natural band widths of flavins (i.e., the width (in nm) of a line joining the two points on the absorbance band of the flavin at which the absorbance is half of the maximum), the choice of appropriate spectral band width for measurements of flavoproteins is relatively straightforward. However, if the band widths chosen are too large, then the observed fluorescence emission may be much lower than the true value. The operator should ensure that the ratio of the spectral band width/natural band width is < 0.2 (preferably < 0.1) to keep the observed signal at close to 100% of its "true" value. Since the natural band width of the longer wavelength flavin absorption band is usually ~ 50–60 nm, this means that the excitation band width (when collecting excitation spectra) or the emission band width (when collecting emission spectra) should preferably be ≤ 5 nm, although the other spectral band width should be greater. Keeping the excitation band width low for emission spectra is also sensible in that it helps to minimize any "photobleaching" of the fluorophore.
3. Fluorescence is much more sensitive to environmental factors than absorption spectrophotometry. Flavoprotein fluorescence experiments should be performed at a constant temperature, since fluorescence emission decreases as temperature increases, due to greater efficiency of quenching in solution *(9)*. If the flavoprotein sample is only weakly fluorescent (as are the flavoprotein components of the mitochondrial respiratory chain, for example), then cooling to sub-zero temperature can significantly improve the signal, with ethylene glycol in the sample as a cryoprotectant. Temperatures down to ~ –40°C can be maintained using a water bath to circulate a water/ethylene glycol mixture around the cuvet. Cooling to

liquid nitrogen temperature may result in up to a 10-fold increase in fluorescence, but requires the use of specialized equipment.

4. Oxygen acts as a quencher and, if signals are particularly low, it may be beneficial to prepare solutions anaerobically. It should also be remembered that oxygen is much more soluble in many organic solvents than it is in water *(9)*. The use of distilled, deionized water for fluorescence measurements will also help to minimize the presence of quenchers in solution.
5. Other precautions that should be made to ensure that signal quality is optimized include
   a. ensuring all faces of the fluorescence cuvet are cleaned with a lens tissue,
   b. thoroughly mixing all samples and removing any drips on the surface of the cuvet,
   c. ensuring that the excitation beam passes through the center of the sample cuvet and not over the meniscus of the sample, and
   d. avoiding turbidity in the sample (which will cause considerable light scattering) by centrifuging the sample briefly or filtering (as described in **Subheading 3.1.1.**).
6. It is unfortunately the case that fluorescence values obtained with one fluorimeter will likely be different to values obtained on another. There is no simple solution to this problem, but methods do exist to "correct" fluorescence spectra for instrumental variation *(46,47)*. A fluorescence excitation spectrum "corrected" for changes in lamp output should be identical to the absorbance spectrum of the same compound *(9)*. However, the Q_f can be used to make reasonable comparisons of fluorescence values from instrument to instrument. For instance, the fluorescence of a known molarity standard solution of FMN ($Q_f = 0.27$) could be determined under identical conditions (the same conditions used to analyze the flavoprotein sample of interest) on different instruments, and the Q_f of the flavoprotein determined by its emission relative to the FMN standard.
7. Several problems may be associated with the fluorimeter used and it is wise to have the machine serviced regularly. As the optical elements of the fluorimeter age, the sensitivity declines. Also, if the machine is miscalibrated there may be shifts observed in the fluorescence maxima. If the scan speed chosen is too fast (or the instrument response is too slow), then a shift is fluorescence emission maximum will be observed *(9)*. If the direction of scanning is from low to high wavelengths, then the emission maximum will be shifted to a higher wavelength, and vice versa.
8. Conclusions

   The fluorescence properties of flavins in proteins helps to define their environments, and characterization of these properties and their perturbation by external factors can be used to monitor such phenomena as protein integrity and flavin exposure, and to measure flavoenzyme reaction kinetics. The fluorescent nature of the flavin isoalloxazine ring system allows protein-bound flavins to be examined without interference from the polypeptide. Fluorescence is thus a key technique in the analysis of flavoproteins.

Acknowledgments

The authors are grateful for the support of the Royal Society of Edinburgh, the Caledonian Research Foundation, the Leverhulme Trust, and the Biotechnology and Biological Sciences Research Council for their research.

References

1. Penzer, G. R. (1980) Molecular emission spectroscopy (fluorescence and phosphorescence), in *An Introduction to Spectroscopy for Biochemists* (Brown, S. B., ed.), Academic Press, London, pp. 70–114.
2. Skoog, D. A. and Leary, J. L. (1992) Molecular fluorescence, phosphorescence, and chemiluminescence spectroscopy, in *Principles of Instrumental Analysis,* 4th ed., Harcourt Brace College Publishers, Orlando, FL, pp. 174–195.
3. Bashford, C. L. (1987) An introduction to spectrophotometry and fluorescence spectrometry, in *Spectrophotometry and Spectrofluorimetry: A Practical Approach* (Harris, D. A. and Bashford, C. L., eds.), IRL Press, Oxford, UK, pp. 1–22.
4. Dawson, R. M. C., Elliot, D. C., Elliot, W. H., and Jones, K. M. (1986) in *Data for Biochemical Research,* 3rd ed., Clarendon Press, Oxford, UK, pp. 124–126.
5. Engel, P. C. (1996) Enzyme cofactors, in *Enzymology Labfax* (Engel, P. C., ed.), Academic Press, San Diego, CA, and Bios Scientific Publishers, Oxford, UK, pp. 223–247.
6. Harvey, R. A. (1980) Flavin 1, N6-ethaneoadenine dinucleotide. *Methods Enzymol.* **66,** 290–294.
7. Mayhew, S. G. and Tollin, G. (1992) General properties of flavodoxins, in *Chemistry and Biochemistry of Flavoenzymes,* vol. III (Müller, F., ed.), CRC Press, Boca Raton, FL, pp. 389–426.
8. Price, N. C. and McLelland, D. A. (1998) Stopped-flow analysis of the refolding of hen egg white riboflavin binding protein in its native and dephosphorylated form. *Biochim. Biophys. Acta* **1382,** 157–166.
9. Poole, R. K. and Bashford, C. L. (1987) Spectra, in *Spectrophotometry and Spectrofluorimetry: A Practical Approach* (Harris, D. A. and Bashford, C. L., eds.), IRL Press, Oxford, UK, pp. 1–22.
10. Kim, J.-J. P., Wang, M., Roberts, D. L., Paschke, R., Shea, T., and Masters, B. S. S. (1997) How do two flavins communicate with each other? The three-dimensional structure of NADPH-cytochrome P450 reductase, in *Flavins and Flavoproteins 1996* (Stevenson, K. J. , Massey, V. and Williams, C. H., Jr., eds.), University of Calgary Press, Calgary, Alberta, Canada, pp. 455–461.
11. Ottada, J., Calcaterra, N. B., Arakaki, A. K., Orellano, E. G., Carrillo, N., and Ceccarelli, E. A. (1997) On the role of aromatic amino acids interacting with FAD in plant-type ferredoxin-$NADP^+$ reductases, in *Flavins and Flavoproteins 1996* (Stevenson, K. J., Massey, V., and Williams, C. H. Jr. P., eds.), University of Calgary Press, Calgary, Alberta, Canada, pp. 501–508.
12. Munro, A. W., Lindsay, J. G., Coggins, J. R., Kelly, S. M., and Price, N. C. (1996) Analysis of the structural stability of the multidomain enzyme flavocytochrome P450 BM3. *Biochim. Biophys. Acta* **1296,** 127–137.

13. Tsou, C.-L. (1986) Location of the active-sites of some enzymes in limited and flexible molecular regions. *Trends Biochem. Sci.* **11,** 427–429.
14. Macheroux, P., Petersen, J., Bornemann, S., Lowe, D. J., and Thorneley, R. N. F. (1996) Binding of the oxidized, reduced and radical flavin species to chorismate synthase—an investigation by spectrophotometry, fluorometry and electron-paramagnetic resonance and electron-nuclear double-resonance spectroscopy. *Biochemistry* **35,** 1643–1652.
15. Giletto, A. and Pace, C. N. (1996) Protein stability, in *Proteins Labfax* (Price, N. C., ed.), Academic Press, San Diego, CA, and Bios Scientific Publishers, Oxford, UK, pp. 233–239.
16. Hlodan, R. and Hartl, F. U. (1994) The protein folding problem, in *Mechanisms of Protein Folding* (Pain, R. H., eds.), IRL Press, Oxford, UK, pp. 194–228.
17. Creighton, T. E. (1994) How the protein folds in the cell, in *Mechanisms of Protein Folding* (Pain, R. H., ed.), IRL Press, Oxford, UK, pp. 1–25.
18. Vermillion, J. L. and Coon, M. J. (1978) Identification of the high and low potential flavins of liver microsomal NADPH-cytochrome P450 reductase. *J. Biol. Chem.* **253,** 8812–8819.
19. Bredt, D. S., Hwang, P. M., Glatt, C. E., Lowenstein, C., Reed, R. R., and Snyder, S. H. (1991) Cloned and expressed nitric oxide synthase structurally resembles cytochrome P-450 reductase. *Nature* **351,** 714–718.
20. Mayhew, S. G. and Wassink, J. H. (1980) Determination of FMN and FAD by fluorescence titration with apoflavodoxin. *Methods Enzymol.* **66,** 217–220.
21. Mayhew, S. G. and Massey, V. (1969) Purification and characterization of flavodoxin from *Peptostreptococcus elsdenii. J. Biol. Chem.* **244,** 794–799.
22. Knight Jr., E. and Hardy, R. F. W. (1966) Isolation and characteristics of flavodoxin from nitrogen-fixing *Clostridium pasteurianum. J. Biol. Chem.* **241,** 2752–2756.
23. Faeder, E. J. and Siegel, L. M. (1973) A rapid micromethod for determination of FMN and FAD in mixtures. *Anal. Biochem.* **53,** 332–336.
24. Weiner, J. H. and Heppel, L. A. (1972) Purification of the membrane-bound and pyridine dinucleotide-independent L-glycerol 3–phosphate dehydrogenase from *Escherichia coli. Biochem. Biophys. Res. Commun.* **47,** 1360–1365.
25. Visser, A. J. W. G., van Hoek, A. V. Kulinski, T., and Le Gall, J. (1987) Time resolved fluorescence studies of flavodoxin. *FEBS Lett.* **224,** 406–410.
26. Visser, A. J. W. G., Ghisla, S., Massey, V., Müller, F., and Veeger, C. (1979) Fluorescence properties of reduced flavins and flavoproteins. *Eur. J. Biochem.* **101,** 13–21.
27. de Kok, A. and Visser, A. J. W. G. (1987) Flavin binding site differences between lipoamide dehydrogenase and glutathione reductase as revealed by static and time-resolved flavin fluorescence. *FEBS Lett.* **218,** 135–138.
28. Visser, A. J. W. G., van Hoek, A., Macheroux, P., and Thorneley, R. N. F. (1997) Time-resolved fluorescence of substrate induced FMN-binding to chorismate synthase, in *Flavins and Flavoproteins 1996* (Stevenson, K. J., Massey, V. and Williams, C. H., Jr., eds.), University of Calgary Press, Calgary, Alberta, Canada, pp. 167–170.

29. Khan, K. K., Mazumdar, S., Modi, S., Sutcliffe, M., Roberts, G. C. K., and Mitra, S. (1997) Steady-state and picosecond-resolved fluorescence studies on the recombinant heme domain of *Bacillus megaterium* cytochrome P-450. *Eur. J. Biochem.* **244,** 361–370.
30. Lehrer, S. S. (1971) Solute perturbation of protein fluorescence: the quenching of the tryptophanyl fluorescence of model compound and of lysozyme by iodide ions. *Biochemistry* **10,** 3254–3263.
31. Ghisla, S., Massey, V., Lhoste, J.-M., and Mayhew, S. G. (1974) Fluorescence and optical characteristics of reduced flavines and flavoproteins. *Biochemistry* **13,** 589–597.
32. Mulrooney, S. B. and Williams Jr., C. H. (1997) Evidence for two conformational states of thioredoxin reductase: Fluorescence studies of treated and untreated C138S mutant, in *Flavins and Flavoproteins 1996* (Stevenson, K. J., Massey, V., and Williams, C. H., Jr., eds.), University of Calgary Press, Calgary, Alberta, Canada, pp. 501–508.
33. van den Berg, P. A. W., van Hoek, A., Visser, A. J. W. G., Walentas, C. D., and Perham, R. N. (1997) Time-resolved flavin fluorescence quenching in *E. coli* glutathione reductase, in *Flavins and Flavoproteins 1996* (Stevenson, K. J., Massey, V., and Williams, C. H., Jr., eds.), University of Calgary Press, Calgary, Alberta, Canada, pp. 501–508.
34. Müller, F., Hopkins, T. R., Lee, J., and Batiens, P. I. H. (1992) Methanol oxidase, in *Chemistry and Biochemistry of Flavoenzymes,* vol. III (Müller, F., ed.), CRC Press, Boca Raton, FL, pp. 96–119.
35. Berry, A., Scrutton, N. S., and Perham, R. N. (1989) Switching kinetic mechanism and putative proton donor by directed mutagenesis of glutathione reductase. *Biochemistry* **28,** 1264–1269.
36. Miller, S. M., Massey, V., Ballou, D. P., Williams Jr., C. H., Moore, M., and Walsh, C. T. (1991) On the mechanism of mercuric reductase: an alternating sites hypothesis, in *Flavins and Flavoproteins 1990* (Curti, B., Ronchi, S., and Zanetti, G., eds.), Walter de Gruyter, Berlin, pp. 627–637.
37. Iyanagi, T. and Mason, H. S. (1973) Some properties of hepatic reduced nicotinamide adenine dinucleotide phosphate-cytochrome *c* reductase. *Biochemistry* **12,** 2297–2308.
38. de Abreu, R. A., de Kok, A., de Graaf-Hess, A. C., and Veeger, C. (1977) Determination of the chain stoichiometries from the number of reactive sulfhydryl groups in the pyruvate dehydrogenase complexes of *Azotobacter vinelandii* and *Escherichia coli*. *Eur. J. Biochem.* **81,** 357–364.
39. Rajagopalan, K. V. and Handler, P. (1972) Nonequivalence of the flavin adenine dinucleotide moieties of chicken liver xanthine dehydrogenase. *J. Biol. Chem.* **247,** 2177–2182.
40. Narayanasami, R., Horowitz, P. M., and Masters, B. S. S. (1995) Flavin-binding and protein structural integrity studies on NADPH-cytochrome P450 reductase are consistent with the presence of distinct domains. *Arch. Biochem. Biophys.* **316,** 267–274.

41. Visser, A. J. W. G., van Hoek, A., Kulinski, T., and Le Gall, J. (1987) Time-resolved fluorescence studies of flavodoxin: demonstration of picosecond fluorescence lifetimes of FMN in *Desulfovibrio* flavodoxins. *FEBS Lett.* **224,** 406–410.
42. Leenders, R., Kooijman, M., van Hoek, A., Veeger, C., and Visser, A. J. W. G. (1993) Flavin dynamics in reduced flavodoxins: a time-resolved polarized fluorescence study. *Eur. J. Biochem.* **211,** 37–45.
43. Bastiaens, P. I. H., van Hoek, A., van Berkel, W. J. H., de Kok, A., and Visser, A. J. W. G. (1992) Molecular relaxation spectroscopy of flavin adenine dinucleotide in wild-type and mutant lipoamide dehydrogenase from *Azotobacter vinelandii. Biochemistry* **31,** 7061–7068.
44. Centeno, F. and Gutierrez-Merino, C. (1992) Location of functional centers in the microsomal P450 system. *Biochemistry* **31,** 8473–8481.
45. Mulrooney, S. B. and Williams, C. H. (1997) Evidence for two conformational states of thioredoxin reductase from *Escherichia coli:* use of intrinsic and extrinsic quenchers of flavin fluorescence as probes to observe domain rotation. *Protein Sci.* **6,** 2188–2195.
46. Parker, C. A. (1968) *Photoluminescence of Solutions.* Elsevier, Amsterdam, The Netherlands.
47. Miller, J. N. (1981) *Standards in Fluorescence Spectroscopy.* Chapman and Hall, London, UK.

4

Potentiometric Measurement of Oxidation-Reduction Potentials

Stephen G. Mayhew

1. Introduction

The free energy change associated with the oxidation or reduction of a purified flavoprotein is assessed by measuring the oxidation-reduction (redox) potential of the bound flavin. The measurement is made by balancing the partially-reduced flavoprotein against another redox system of known redox potential *(1,2)*. The reference can be a second redox system in solution, such as a well-characterized dye, in which case the flavoprotein and dye are brought into an equilibrium that can be analyzed spectroscopically. Alternatively, the potential in the solution of partially reduced flavoprotein is balanced against the potential of a standard electro chemical half-cell, such as a mercury/mercuric chloride (calomel) electrode. In the second case, the redox potential of the flavoprotein is probed with an inert metal such as gold or platinum, electrical connection between the flavoprotein and the calomel electrode is completed with a salt bridge made from saturated KCl, and the difference in potential between the partially reduced flavoprotein and the standard electrochemical half-cell is determined with a voltmeter. A value for the redox potential at half-reduction of the flavoprotein is calculated (E_M) and expressed in volts with the standard hydrogen electrode as the point of reference ($E_M = 0$ V). When E_M for one system is less positive (or more negative) than E_M for a second system, the first system is the more reducing and the direction of electron transfer will tend to be towards the second system.

The redox potentials for the 2-electron reductions of flavin mononucleotide (FMN) and flavin adenine dinucleolide (FAD) in free solution at pH 7 are –0.205 V *(3)* and –0.219 V *(4)*, respectively. However, since a small amount of the flavin semiquinone is formed in thermodynamic equilibrium with the quinone and

From: Methods in Molecular Biology, vol. 131: Flavoprotein Protocols
Edited by: S. K. Chapman and G. A. Reid © Humana Press Inc., Totowa, NJ

hydroquinone (**Eq. 1**), the overall 2-electron reduction can be analyzed for the two 1-electron transfer steps, and a redox potential assigned to each step (**Eqs. 2** and **3**):

$$2(\text{semiquinone}) \leftrightarrow \text{quinone} + \text{hydroquinone} \quad (1)$$

$$\text{quinone} + 1e^- \leftrightarrow \text{semiquinone} \quad (2)$$

$$\text{semiquinone} + 1e^- \leftrightarrow \text{hydroquinone} \quad (3)$$

In the case of FMN and riboflavin, the semiquinone formed at half reduction of the flavin comprises only about 10% of the total flavin in solution *(3)*. Theory shows *(1)* that when the equilibrium of Eq. 1 lies so far to the right, E_M for reduction of the quinone to the semiquinone is actually more negative than E_M for the reduction of semiquinone to the hydroquinone (–0.238 V and –0.172 V respectively, for FMN at pH 7 *(3)*).

The redox potentials of protein-bound flavin vary widely *(5)*. In addition, the affinity of the protein for the three redox states of the flavin may differ, and the protein may alter the ionization of the flavin. For example, the semiquinone is thermodynamically stabilized in many flavoproteins, and the pK_a at 8.5 associated with the semiquinone in free flavin may be increased or decreased *(5)*. Preferential stabilization of the semiquinone leads to a greater separation of the redox potentials of the two steps in reduction *(1)*. Measurement of the redox potentials of the bound flavin, and of any other redox groups associated with a flavoprotein, can provide information about the relevance of the different redox forms to the overall reaction. Detailed analyses of the effects of pH and of ligands, such as substrates and inhibitors, on the redox potentials have been carried out with a few flavoenzymes *(5)*. With complementary information about the structure of the flavin-binding site, the strength of interaction of the flavin and protein, and about the effects of changing these interactions chemically or by site-directed mutagenesis, an understanding of the forces that regulate flavoprotein redox potentials is beginning to emerge.

The methods that have been used to measure the redox potentials of flavoenzymes vary in their requirements for specialized equipment and in the ease with which they can be applied to different flavoenzymes. Optical analysis of the equilibrium formed between partially-reduced flavoenzyme and a similar concentration of a reference redox system of known midpoint potential, as mentioned earlier, is perhaps the simplest method. It requires only a dye with a redox potential similar to that of the flavoprotein, a spectrophotometer, a cuvet that can be made anaerobic, and a method to introduce reducing equivalents into the mixture, preferably in a stepwise manner.

$$\text{flavoprotein}_{ox} + \text{reference dye}_{red} = \text{flavoprotein}_{red} + \text{reference dye}_{ox} \quad (4)$$

The reference system allows the redox potential (E_h) in the solution to be determined from the Nernst equation, and the midpoint potential of the flavoenzyme can then be determined from the extent of reduction of the flavin measured in the equilibrium. Detailed prior knowledge of any redox intermediates and of the changes in the absorption spectrum of the flavoenzyme during the reduction is required in this method because most dyes absorb light in the same region as flavin. In addition, if the method is to be used to investigate the effects of changing an experimental condition, such as pH, the effect of the change on the spectrum of the dye must be determined independently. Massey *(6)* described an interesting variation of this method in which uric acid and a catalytic concentration of xanthine oxidase is used to effect the slow reduction of a mixture of a redox dye and a flavoprotein. Optical measurements during the course of the reduction are analyzed for the concentrations of oxidized and reduced dye, and the concentrations of oxidized and reduced flavoprotein. It is assumed that reduction is sufficiently slow to allow the further assumption that throughout the reaction the flavin in the enzyme is in redox equilibrium with the dye. A value for the midpoint potential of the flavoenzyme can then be calculated at different time intervals. The method was validated with flavoenzymes of known redox potential.

Spectroelectrochemical methods which combine potentiometric measurement of the redox potential in a solution of the protein with optical measurement of the redox state of the bound flavin are more versatile. In addition, although a redox dye is usually included in the solution to mediate electron transfer between the protein-bound flavin and the metal electrode that is used to sense the potential in the solution, the dye is present at a much lower concentration than the enzyme, and therefore the dye does not obscure the absorption spectrum. One such method is described in **Subheadings 2.–4.** In this method, the potential in solution that is sensed by a gold electrode is determined by balancing it against the potential of a standard calomel electrode. In its simplest form, the method requires very little equipment that is not available in the general biochemical laboratory.

As with any method that involves the quantitative and partial reduction of a flavoprotein, it is essential to maintain the enzyme completely free of oxygen during the long periods often required for equilibration of the system. This can be achieved most effectively by enclosing the measuring equipment in an anaerobic glove box. However, such a specialized facility is not usually available, and therefore the method described in **Subheading 3.1.** uses an all-glass cuvet that is made anaerobic by connection to a vacuum pump and to a supply of purified nitrogen gas. Continuous flushing of the cuvet with nitrogen during the measurement is avoided in order to minimize the effects of trace amounts of oxygen that may still contaminate the nitrogen.

The method uses chemical reduction with sodium dithionite to introduce reducing equivalents into the flavoprotein because dithionite ion is a strong reducing agent that is widely available *(7)*. However, dithionite ion suffers the disadvantage that in solution it reacts readily with oxygen, so that special techniques are required for the preparation and use of solutions of its salt. In addition the oxidation product of dithionite, (bi)sulphite, forms an adduct with the oxidized flavin of certain flavoenzymes *(8)*. A method that provides a technically less complex procedure for introducing the reducing equivalents in a controlled way is the use of photoreduction with a 5-deazaflavin *(9)*. Unfortunately this flavin analogue is not available commercially, and the use of other photocatalysts does not usually lead to full reduction of a flavoenzyme. Electrochemical reduction is a similarly convenient method of reducing a flavoprotein in a controlled way *(10)*, but the method requires a potentiostat or similar device not often available in a biochemical laboratory. The anerobic cuvet shown in **Fig. 1** and described in **Subheading 2.** for use with dithionite titration can be readily adapted to methods that use either photochemical or electrochemical reduction. It also allows the chemical oxidation of a reduced flavoenzyme by titration of the enzyme with ferricyanide ion, a strong oxidant.

2. Materials

1. An all-glass cuvet (modified from a Starna anerobic cuvet, 128/SOG/10 [Starna Cells Inc., Atascadero, CA]), similar to that of **Fig. 1**, fitted with sockets to accommodate a tap (through which the cuvet is evacuated and filled with nitrogen) and two electrodes, and a cone to take a syringe (to deliver reductant or oxidant). The cuvet is adapted from cuvets described elsewhere *(7,10–13)*. The cuvet is equipped with a side arm that contains photoreduced methyl viologen to react with any contaminating oxygen.
2. A gold foil electrode. This can be made by heating pieces of gold foil and gold wire to red heat and then fusing them together with a sharp blow from a hammer. Using a small amount of epoxy resin, the wire is fixed into a glass tube that is fitted with a glass cone (**Fig. 1**).
3. A salt bridge. This is made from a tube fitted with a cone and a socket joined in-line. The tube is drawn to a fine point and filled with agar in saturated KCl.
4. A standard calomel electrode (microelectrode, 6 mm diameter). The glass tip with its porous plug is replaced by a glass cone, and the modified electrode fitted into the socket of the salt bridge. Contact between the electrode and agar-KCl of the bridge is made with saturated KCl that is added through the usual filling hole on the electrode.
5. A suitable buffer, such as 50 m*M* sodium phosphate, pH 7.
6. A range of dyes of different redox potential *(1)*, added at a concentration of 1 μ*M* to mediate electron transfer between the protein-bound flavin and the gold electrode. The following are useful with flavoenzymes (redox potential at pH 7 and

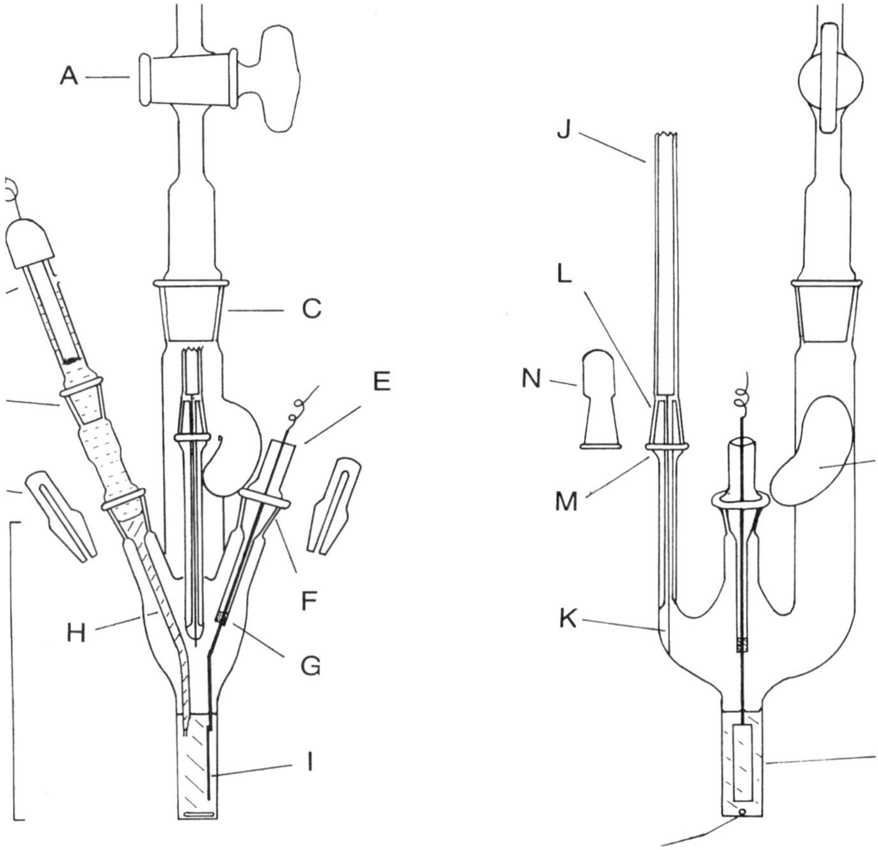

Fig. 1. Anaerobic cuvet. Left: front view. Right: side view with calomel electrode omitted for clarity. (A) Vacuum tap joined to 14/23 cone. (B) Calomel electrode (micro-electrode, 6 mm diameter). (C) 14/23 Socket. (D) 14/23 Cone closed at end to form a cap. (E) 14/23 Cone joined to glass tube. (F) 14/23 Sockets. (G) Plug formed from epoxy cement. (H) Salt bridge made by joining a 14/23 socket (F) to a 14/23 cone that is extended with a glass tube to a narrow tip that dips into the solution in the cuvet; the glass tube is filled with molten agar that is saturated with KCl (40*g* KCl in 100 mL boiling agar, 3% (w/v)); after it has cooled and set, the agar is overlayed with saturated KCl to complete electrical contact between the calomel electrode and the test solution in the cuvet. (I) Gold electrode made by heat-fusing gold foil (20 mm × 4 mm × 0.1 mm) to a gold wire (100 mm × 0.5 mm diam.); the wire is sealed into the glass tube with a plug of epoxy cement (G). (J) Hamilton Gas-Tight threaded plunger syringe (1750 TPLT), 0.5 mL, fitted with an 8.5 cm needle (K) and joined to a 14/23 socket (L) with epoxy cement. (M) 14/23 Cone joined to a narrow-bore tube. (N) 14/23 socket closed at end to form a cap. (O) Side arm that contains 0.3 mL methyl viologen mixture. (P) Glass cuvet (modified anaerobic Starna cell 128/SOG/10). (Q) Magnetic flea.

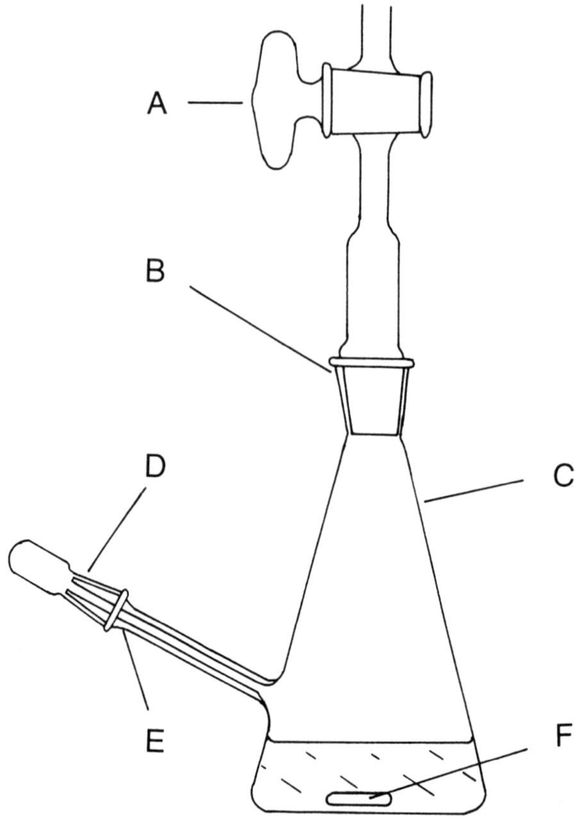

Fig. 2. Titrant-preparation flask. (A) Vacuum tap joined to a 14/23 cone. (B) 14/23 Socket. (C) 125 mL Erlenmeyer flask. (D) 14/23 Socket closed at end to form a cap. (E) 14/23 Cone joined to narrow-bore tube. (F) Magnetic flea.

25°C given in brackets): methyl viologen (–0.446 V); benzyl viologen (–0.359 V); neutral red (-0.325 V); safranine T (–0.289 V); phenosafranine (–0.252 V); anthraquinone 2-sulfonate (–0.225 V); anthraquinone 2,6 disulfonate (–0.184 V); 2-OH-1,4-naphthoquinone (–0.152 V); indigo disulfonate (–0.116 V); indigo tetrasulfonate (–0.046 V); pyocyanine (–0.038 V); methylene blue (0.008 V); phenazine ethosulfate (0.055); phenazine methosulfate (+ 0.08 V). The viologens are 1-electron carriers and their use can facilitate formation of the flavin semiquinone.

7. A titrant-preparation flask (**Fig. 2**).
8. A manifold through which the cuvet can be alternately evacuated at a vacuum pump and filled with oxygen-free nitrogen (**Fig. 3**). Nitrogen gas from the cylinder is freed of oxygen by passing it through a column of catalyst at 120°C (BASF catalyst, R3-11, BASF Corp., Mount Olive, NJ) and saturated with water vapor in a wash bottle of reduced methyl viologen.

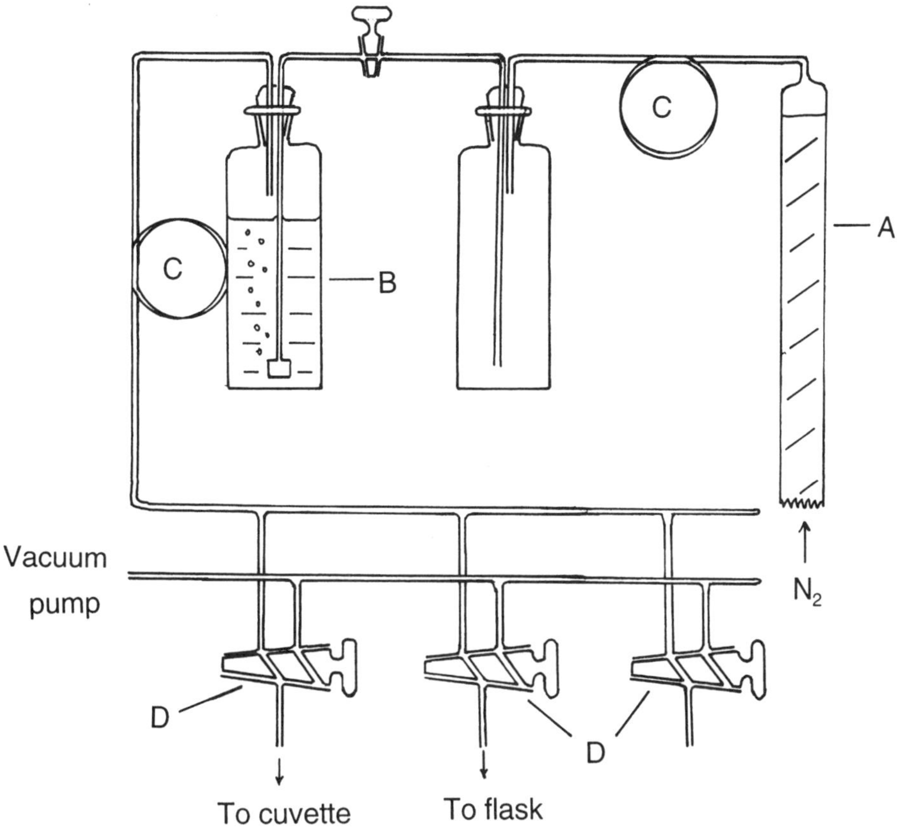

Fig. 3. Schematic arrangement of gas and vacuum line connections. (A) Glass column (80 cm × 3 cm diam.) two-thirds filled with BASF catalyst (R3-11) and heated to 120°C with a heating tape. (B) 250 mL Wash-bottle containing 150 mL methyl viologen mixture. (C) Coils of 0.125 inch copper tubing sealed to glass tubing with epoxy cement. (D) Three-way Vacuum taps; connections to cuvet and titrant-preparation flask are made with butyl rubber tubing (0.25 in. internal diameter, 0.125 in. wall thickness).

9. A scanning spectrophotometer fitted with a thermostatted cell housing at 25°C and a magnetic stirrer.
10. A voltmeter capable of measuring ± 1 V to ± 0.1 mV (a laboratory pH meter may be adequate), and preferably equipped to provide a continuous record. One pole of the meter is connected to the output from the calomel electrode; the other is connected to the gold electrode.
11. Apiezon N stopcock grease (Apiezon Products, Manchester, UK) for all taps and ground-glass joints.
12. Methyl viologen mixture used in the side arm of the cuvet and in the wash bottle in the nitrogen supply line (**Fig. 3**): 1 m*M* methyl viologen; 3 µ*M* proflavin;

0.1 *M* Tris-HCl buffer, pH 8; and 30 m*M* EDTA. After making the solution anaerobic, the methyl viologen is exposed to light until it is partly reduced to the semiquinone.

3. Methods

3.1. Preparation of Cuvet

1. Clamp the cuvet on a magnetic stirrer adjacent to the vacuum/gas line.
2. Add to the main compartment of the cuvet in a final volume of 3.2 mL: flavoenzyme (20–60 μ*M* flavin, final concentration), buffer, (a) redox dye(s) (1 μ*M*) (*see* **Note 1**), and a magnetic flea.
3. Add 0.3 mL of the methyl viologen mixture to the side arm on the cuvet.
4. Place caps in the two electrode sockets and over the cone for the syringe.
5. Place the tap on the cuvet and connect the tap to a three-way tap on the vacuum/gas line (**Fig. 3**) using a short length of butyl rubber tubing. Switch on the magnetic stirrer.
6. Make the cuvet and contents anaerobic by alternate evacuation with the vacuum pump and filling with nitrogen gas (6 cycles during about 20 min). Maintain the optical part of the cuvet and the cuvet contents at 4°C in ice-water during this process, and stir the contents continuously using the magnetic stirrer.

3.2. Preparation of Sodium Dithionite

The solution of sodium dithionite is prepared in the titrant flask.

1. Clamp the flask on a magnetic stirrer adjacent to the vacuum/gas line, and place a cap over the side arm on the flask.
2. Add to the flask, 45 mL 0.1 *M* sodium pyrophosphate-HCl buffer, pH 9, and a magnetic flea.
3. Place the tap on the flask and connect it to a second 3-way tap on the gas/vacuum line (**Fig. 3**) using a short length of butyl rubber tubing. Switch on the magnetic stirrer.
4. Make the flask and contents anaerobic as described in **Subheading 3.1.**, step 6.
5. Switch off the magnetic stirrer, and open the flask by removing the tap. Quickly add 25 mg solid sodium dithionite to the flask using a glass funnel to direct the crystalline powder away from the grease in the neck of the flask, and allow it to sink to the bottom (*see* **Note 2**). Close the flask immediately, replace the tap, and evacuate and fill the flask with nitrogen (3 cycles). Turn on the stirrer for a few seconds to mix the flask contents.
6. Remove the cap on the side arm to the flask, allowing nitrogen to flow through the flask. Insert the needle of the modified Hamilton syringe (Hamilton Co., Reno, NV) loosely into the side arm and operate the plunger several times to expel air from the barrel and needle into the stream of nitrogen. Then push the socket on the syringe firmly onto the cone to stop the flow of gas, and fill the syringe with dithionite solution. Remove the filled syringe and replace the cap on the side arm

before reevacuating the flask. Dithionite prepared and stored in this way is stable for several days at 20°C *(7)*.

3.3. Assembly of the Electrodes and Syringe on the Cuvet

1. Remove the cap that covers the cone on the cuvet allowing nitrogen to flow through the cuvet and to exit via the cone. Place the filled syringe onto the cone, allowing gas to escape between the cone and socket for a few seconds before pushing the socket firmly into place.
2. Remove one of the caps that cover the two electrode ports on the cuvet, allowing nitrogen to flow through the socket. Quickly insert the fitting that contains the salt bridge and calomel electrode, and allow nitrogen to flow between the socket and cone for a few seconds before pushing the fitting home.
3. Remove the cap that covers the second electrode port and insert the gold electrode under a flow of nitrogen as described above.
4. Open the third 3-way tap on the vacuum/gas manifold momentarily while at the same time closing the tap on the cuvet; this brings the pressure in the cuvet to atmospheric pressure as the cuvet is sealed.
5. Disconnect the cuvet from the vacuum/nitrogen line. Taking care to protect the enzyme from light, shine the light from a slide projector (250 W tungsten-halogen bulb, Kodak Carousel S-AV2000, Kodak, Rochester, NY) onto the side arm of the cuvet to partially convert the methyl viologen to its blue reduced form. Gently shake the cuvet for several minutes to allow any oxygen that remains in the cuvet to react with the reduced methyl viologen.
6. Place the cuvet into the cell holder of the spectrophotometer thermostatted at 25°C, switch on the magnetic stirrer in the spectrophotometer, and connect the calomel and gold electrodes to the input sockets on the millivoltmeter.
7. After recording a spectrum, add dithionite solution from the syringe (5.29 μL per turn), and mix it with the enzyme by tipping the cuvet contents over the tip of the syringe needle. Monitor the optical spectrum and the potential until equilibrium is attained or until the change in potential is less than 0.1 mV/10 min (*see* **Note 3**). Record the spectrum before making a further addition of dithionite.
8. When reduction of the enzyme is complete (usually at least 10 additions of dithionite), an oxidative titration can be carried out with ferricyanide ion. Reconnect the cuvet to the vacuum/gas manifold before removing the syringe under a flow of nitrogen, and replacing it with a syringe filled with a solution of potassium ferricyanide (6 m*M*) made anaerobic in a titrant flask as described in **Subheading 3.2.** Using the procedures outlined in steps 6 and 7, monitor the optical spectrum and potential during stepwise oxidation of the enzyme (*see* **Note 4**).

3.4. Data Analysis

The optical spectra are analyzed at each step of the titration to determine the proportions of flavoprotein in the relevant oxidized and reduced forms. Where necessary, corrections are made for the absorbance of the redox dye(s) added as mediator(s). The observed potential in volts at each step is corrected for the potential of the calomel electrode (0.2444 V at 25°C, *(1,2)*). The potential at 50% reduction of the flavoenzyme is determined using the Nernst equation:

$$E_h = E_M + 2.303(RT/nF)\log_{10}([ox]/[red]) \quad (4)$$

where E_h is the observed potential vs the standard hydrogen electrode; E_M is the potential at 50% reduction (the midpoint potential when concentrations of the oxidized and reduced forms of the redox couple are equal); R is the gas constant (8.31 J K^{-1} mol^{-1}); T is the temperature in K; n is the number of electrons to convert *ox* into *red*; F is the Faraday (96.5 kJ V^{-1} mol^{-1}); and *ox* and *red* are the relevant oxidized and reduced forms of the flavoprotein (they could be the semiquinone and hydroquinone forms of the flavin if the 1-electron reduction of the semiquinone is under study).

A plot is made of E_h vs $\log_{10}[ox]/[red]$, and the value for E_M determined from the point of intersection of the resulting line with the y-axis when $\log_{10}[ox]/[red]$, is zero. The slope ($2.303 \log_{10}RT/nF$) has the value 0.0296 V at 25°C when $n = 2$, and 0.0592 V when $n = 1$.

4. Notes

1. The redox potential of the dye should be similar (± 0.06 V) to that of the redox couple of the flavoenzyme. The choice of dye is arbitrary until the redox potential of the flavoenzyme is known, and it may be convenient to use several dyes together in initial experiments *(1,2)*. At least two dyes will usually be required if measurements on the two 1-electron steps in reduction of the flavin successively to the semiquinone and hydroquinone are to be determined in the same experiment.
2. The dithionite solution should be about 3 m*M*. Note that sodium dithionite is usually about 85% pure when freshly purchased, but that the dry solid becomes oxidized in air.
3. The time required to attain equilibrium is variable. At least 30 min may be required at the start of a reductive titration. The time required in oxidative titrations is usually shorter than in reductive titrations.
4. The data from the oxidative titration should coincide with those from the reductive titration.

References

1. Clarke, W. M. (1960) *Oxidation-Reduction Potentials of Organic Systems,* Williams and Wilkins, Baltimore, MD, 584 pp.
2. Dutton, P. L. (1978) Redox potentiometry: determination of midpoint potentials of oxidation-eduction components of biological electron-transfer systems, in *Methods in Enzymology* (Fleischer, S. and Packer, L., eds) vol. 54, Academic Press, New York, pp. 411–435.
3. Draper, R. D. and Ingraham, L. L. (1968) A potentiometric study of flavin semiquinone equilibrium. *Arch. Biochem. Biophys.* **125,** 802–808.

4. Lowe, H. J. and Clarke, W. M. (1956) Studies on oxidation-reduction. XXIV. Oxidation-reduction potentials of flavin adenine dinucleotide. *J. Biol. Chem.* **221,** 983–992.
5. Stankovich, M. T. (1991) Redox properties of flavins and flavoproteins, in *Chemistry and Biochemistry of Flavoenzymes* (Müller, F., ed.), vol 1, CRC Press, Inc., Boca Raton, FL, pp. 401–425.
6. Massey, V. (1991) A simple method for determination of redox potentials, in Flavins and Flavoproteins (Curti, B., Ronchi, S., and Zanetti, G., eds.), Walter de Gruyter, Berlin, pp. 59–66.
7. Mayhew, S. G. (1978) The redox potential of dithionite and SO_2^- from equilibrium reactions with flavodoxins, methyl viologen and hydrogen plus hydrogenase. *Eur. J. Biochem.* **85,** 535–547.
8. Massey, V., Müller, F., Feldberg, R., Schuman, M., Sullivan, P. A., Howell, L. G., Mayhew, S. G., Matthews, R. G., and Foust, G. P. (1969) The reactivity of flavoproteins with sulfite. Possible relevance to the problem of oxygen reactivity. *J. Biol. Chem.* **244,** 3999–4006.
9. Massey, V. and Hemmerich, P. (1978) Photoreduction of flavoproteins and other biological compounds catalyzed by deazaflavins. *Biochemistry* **17,** 9–16.
10. Stankovich, M. (1980) An anaerobic spectroelectrochemical cell for studying the spectral and redox properties of flavoproteins. *Anal. Biochem.* **109,** 295–308.
11. Burleigh, D. B. Jr., Foust, G. P., and Williams, C. H., Jr. (1969) A method for titrating oxygen-sensitive organic redox systems with reducing agents in solution. *Anal. Biochem.* **27,** 536–544.
12. Foust, G. P., Burleigh, D. B., Jr., Mayhew, S. G., Williams, C. H., Jr., and Massey, V. (1969) An anaerobic titration assembly for spectrophotometric use. *Anal. Biochem.* **27,** 530–535.
13. Lambeth, D. O. and Palmer, G. (1973) The kinetics and mechanism of reduction of electron transfer proteins and other compounds of biological interest by dithionite. *J. Biol. Chem.* **248,** 6095–6103.

5

Flavoprotein Kinetics

Willem J. H. van Berkel, Jacques A. E. Benen, Michel H. M. Eppink, and Marco W. Fraaije

1. Introduction

Flavoproteins are ubiquitous proteins involved in diverse biological processes ranging from redox catalysis and light emission to DNA repair *(1)*. Based on their function, flavoproteins can be divided into several subclasses including electron transferases, photolyases, synthases, dehydrogenases, disulfide reductases, oxidases, and monooxygenases. Most flavoproteins contain a single flavin adenine dinucleotide (FAD) or flavin mononucleotide (FMN) molecule as noncovalently bound prosthetic group. Some flavoproteins, like reduced nicotinamide adenine dinucleotide phosphate (NADPH) cytochrome P-450 reductase, contain both FAD and FMN but there are also many flavoproteins that harbor different types of cofactors. Examples of these more "complex" flavoproteins include xanthine oxidase (containing molybdenum and 2Fe-2S centres), trimethylamine dehydrogenase (containing a 4Fe-4S cluster) and flavocytochromes (containing heme).

Because of their unique optical properties, flavoproteins can be conveniently studied by spectroscopic techniques (*see* Chapter 1). In this chapter we will describe "kinetic" protocols for three major classes of flavoenzymes of which the catalytic mechanism has been well-studied: disulfide reductases (*see* **Subheading 1.1.**), oxidases (*see* **Subheading 1.2.**) and monooxygenases (*see* **Subheading 1.3.**).

1.1. Disulfide Reductases

NAD(P)H-dependent FAD-containing disulfide reductases are homologous homodimeric enzymes with diverse biological functions *(2)*. The active sites of these flavoenzymes are located at the dimer interface and residues from both

From: Methods in Molecular Biology, vol. 131: Flavoprotein Protocols
Edited by: S. K. Chapman and G. A. Reid © Humana Press Inc., Totowa, NJ

subunits participate in catalysis ***(3)***. Examples of flavoprotein disulfide reductases of which the kinetic mechanism has been extensively studied are: glutathione reductase ***(4)***, mercuric ion reductase ***(5,6)***, thioredoxin reductase ***(7–9)***, and (dihydro)lipoamide dehydrogenase ***(10–15)***. Based on properties of lipoamide dehydrogenase from *Azotobacter vinelandii*, we here describe a general strategy to study the catalytic mechanism of disulfide reductases.

Lipoamide dehydrogenase (LipDH, EC 1.8.1.4) catalyses the oxidation by NAD^+ of a dihydrolipoyl group which is covalently bound to the dihydrolipoyl transferase core component of the 2-oxoacid dehydrogenase multienzyme complexes ***(16,17)***. The flow of electrons is from the dihydrolipoyl group to the active-site disulfide, from the protein dithiol to FAD and from the reduced FAD to NAD^+. The enzyme is also active with free dihydrolipoamide ($lip(SH)_2$), facilitating kinetic studies:

$$\mathrm{lip(SH)_2 + NAD^+ \underset{}{\overset{E\text{-}FAD}{\rightleftharpoons}} lipS_2 + NADH + H^+} \tag{1}$$

Three basic types of LipDH assays have been described, with various modifications ***(18)***. The most frequently used method concerns the "reverse" reaction, i.e., the NADH-dependent reduction of lipoamide ($lipS_2$). However, this reaction is ionic strength dependent and requires the presence of NAD^+ in order to prevent dead-end inhibition ***(13,19,20)***. Another frequently cited assay concerns the so-called diaphorase activity, i.e., the NADH-dependent reduction of 2,6-dichlorophenolindophenol ***(21)***. However, this reaction is not strictly related to the number of intact active sites ***(18,22)***. Therefore, we recommend as standard LipDH assay the NAD^+-dependent oxidation of $lip(SH)_2$ ***(23)***, i.e., the "physiological" reaction depicted in **Eq. 1**.

Steady-state kinetics of LipDH are preferably performed at pH 8.0, the optimum pH for the $lip(SH)_2/NAD^+$ reaction ***(2)***. From varying the concentration of the first substrate at a fixed concentration of the second substrate, and treating the data according to Michaelis-Menten kinetics, parallel lines are obtained in double-reciprocal plots, indicative of a ping-pong mechanism ***(15)***. This shows that *A. vinelandii* LipDH acts according to the following sequence of events:

$$\mathrm{E_{ox} + lip(SH)_2 \underset{k_{-1}}{\overset{k_1}{\rightleftharpoons}} E_{ox}lip(SH)_2 \underset{k_{-2}}{\overset{k_2}{\rightleftharpoons}} EH_2 + lipS_2} \tag{2}$$

$$\mathrm{EH_2 + NAD^+ \underset{k_{-3}}{\overset{k_3}{\rightleftharpoons}} EH_2NAD^+ \underset{k_{-4}}{\overset{k_4}{\rightleftharpoons}} E_{ox} + NADH + H^+} \tag{3}$$

where E_{ox} and EH_2 represent the oxidized and two-electron reduced enzyme, respectively. According to these reactions, EH_2 is a free "unliganded" enzyme species. EH_2 can also be obtained by reduction of E_{ox} by dithionite or borohydride. This allows the kinetics of the reductive (**Eq. 2**) and oxidative half-reactions (**Eq. 3**) to be studied independently by rapid-reaction techniques. For this purpose, it is essential to know the optical properties of the enzyme in the various redox states.

The presence of two redox-active groups in the active site, the FAD and the disulfide, imply that LipDH can take up four reducing equivalents. However, it has been well established that during turnover, the enzyme shuttles between E_{ox} and EH_2 *(23)*. **Figure 1** shows the spectral characteristics of *A. vinelandii* LipDH at pH 8.0 *(24,25)*. In the oxidized state, the yellow protein exhibits absorption maxima around 360 and 455 nm. Upon reduction by lip$(SH)_2$, the enzyme turns red and a typical absorbance arises around 530 nm due to the formation of a charge transfer interaction between the oxidized flavin and a protein thiolate *(23)*. The intensity of this charge transfer absorption band is strongly pH dependent as influenced by the pK_a modulating properties of a C-terminal histidine which acts as an active-site base *(2,24,26)*. Under anerobic conditions, the red EH_2 species is stable in the presence of a small excess of lip$(SH)_2$, showing that the reducing equivalents mainly reside on the disulfide. Reduction of E_{ox} by NADH leads to a large absorbance decrease at 450 nm and the formation of a broad absorption band around 700 nm (**Fig. 1**). The absorption spectrum of the resulting grey species is indicative of the stabilization of a charge transfer complex between the reduced flavin and NAD^+ *(2,15,23)*. Furthermore, it has been shown that binding of NAD^+ to EH_2 shifts the reducing equivalents from the disulfide towards the flavin due to an increase of the redox potential of the FAD *(12,15)*.

1.2. Flavoprotein Oxidases

Biological oxidation reactions mostly involve the rupture of a C-H bond with a concomitant transfer of two electrons to an electron acceptor like $NAD(P)^+$, cytochrome, or FAD. Reactions catalyzed by flavoprotein oxidases generally involve two substrates, an electron donor (substrate) and molecular oxygen acting as the electron acceptor:

$$S + O_2 \xrightarrow{\text{E-FAD}} P + H_2O_2 \tag{4}$$

Therefore, these enzymatic reactions can be characterized as two-substrate two-product reactions. As a consequence, most flavoprotein oxidases obey either a ping-pong mechanism or a ternary complex mechanism. The type of kinetic mechanism varies between different flavoprotein oxidases and can depend on the type of substrate. Kinetic analysis of the different enzymes has revealed that the rate-limiting step in catalysis is often represented by either the rate of flavin reduction or the rate of product release. As for the type of mechanism, the rate-limiting step may vary depending on the type of substrate.

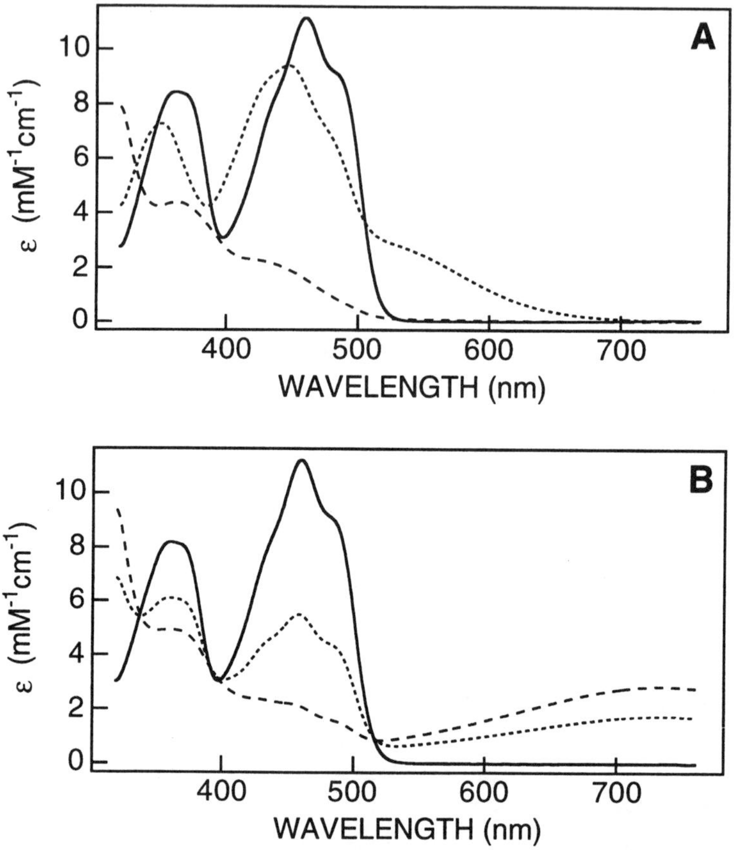

Fig. 1. Optical properties of lipoamide dehydrogenase from *Azotobacter vinelandii*. The experiments were performed under anaerobic conditions at 25°C, pH 8.0 ***(25)***. (A) Visible absorption spectrum of oxidized LipDH (-) and after reduction with borohydride to EH_2 (·····) or with dithiothreitol to EH_4 (- - -). The charge-transfer absorbance around 550 nm of EH_2 indicates that the reducing equivalents reside on the disulfide. (B) Visible absorption spectrum of oxidized LipDH (-) and after reduction with one (·····) or two (- - -) molar equivalents of NADH. The absorbance around 730 nm is indicative for a charge transfer interaction between reduced flavin and NAD^+.

Flavoprotein oxidase catalysis involves two half-reactions in which first the flavin is reduced by the substrate (reductive half-reaction) and subsequently the reduced flavin is reoxidized by molecular oxygen (oxidative half-reaction). As oxidized and reduced flavins differ in their spectroscopic properties (absorbance, fluorescence), the transitions between the oxidized and reduced state can conveniently be monitored. Based on this, the two half-reactions can be analyzed separately using the stopped-flow technique. These features have permitted detailed kinetic studies on several flavin-dependent oxidases. Examples of flavoprotein oxidases of which the kinetic mechanism has been extensively studied are: D-amino acid oxidase *(27)*, L-lactate oxidase *(28)*, monoamine oxidase *(29)*, and vanillyl-alcohol oxidase *(30,31)*. Based on the properties of the latter enzyme, we here describe a general strategy to study the catalytic mechanism of flavoprotein oxidases.

Vanillyl-alcohol oxidase (VAO, EC 1.1.3.13) is a covalent flavoprotein isolated from the fungus *Penicillium simplicissimum* *(32)*. The enzyme is active with a wide range of phenolic compounds *(31,33)*. The biological function of VAO is unclear but induction experiments have shown that the enzyme catalyses the first step in the biodegradation of 4-(methoxymethyl)phenol *(34)*. A detailed kinetic study with this compound revealed that the enzymatic reaction involves the formation of a binary complex of reduced VAO and the *p*-quinone methide of 4-(methoxymethyl)phenol *(30)*. This complex then reacts with molecular oxygen, reoxidizing the flavin, and after water addition to the *p*-quinone methide, the products 4-hydroxybenzaldehyde, and methanol are released:

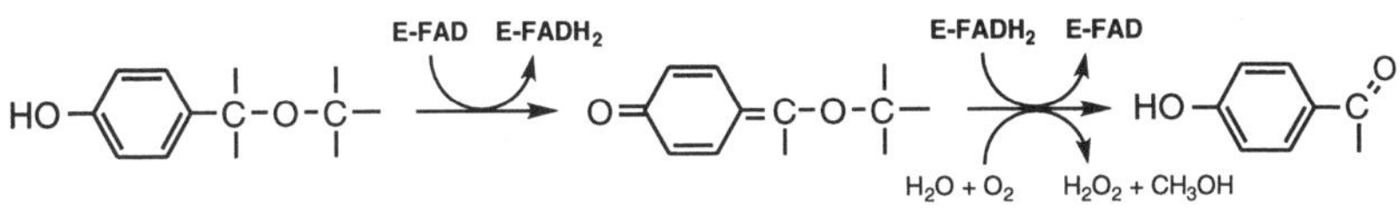

(5)

VAO is a homo-octamer with each 65 kDa subunit containing an 8α-(N^3-histidyl)-FAD *(32)*. Elucidation of the crystal structure revealed that the VAO subunit consists of two domains with the active site being located at the domain interface *(35)*. Furthermore, sequence alignments have indicated that the larger domain represents a novel type of FAD-binding fold which favors covalent flavinylation *(36)*.

The activity of flavoprotein oxidases can be determined by measuring the rate of oxygen consumption using an oxygen electrode. However, for VAO we recommend as standard assay the spectrophotometric determination of vanillin production from vanillyl alcohol which can be conveniently monitored at 340 nm *(33)*. With the majority of VAO substrates, the pH optimum of catalysis is around pH 9.5. However, steady-state kinetics have mostly been performed at pH 7.5. For the reaction with 4-(methoxymethyl)phenol, and by varying the concentration of the first substrate at a fixed concentration of the second substrate, parallel lines are observed in double-reciprocal plots, indicative of a ping-pong mechanism *(30)*. However, from measuring the reductive and oxidative half-reactions by stopped-flow kinetics, evidence was obtained that with this substrate, VAO

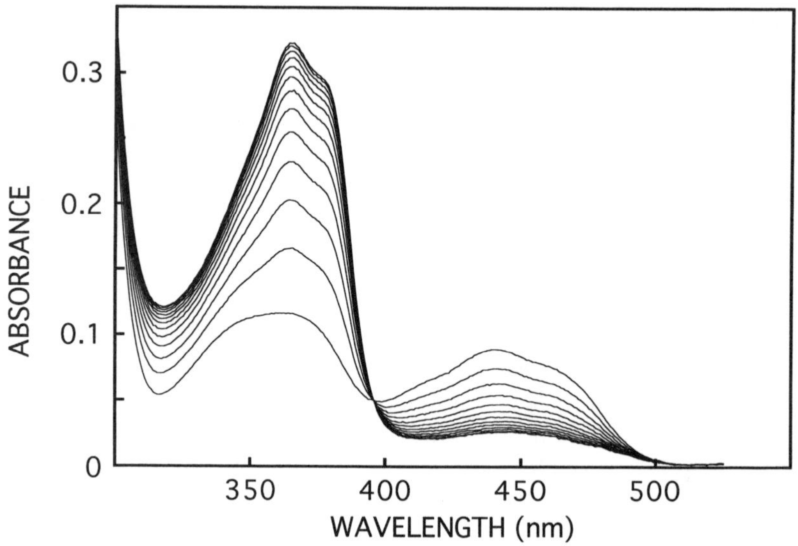

Fig. 2. Reductive half-reaction of vanillyl-alcohol oxidase from *Penicillium simplicissimum* with 4-(methoxymethyl)phenol. VAO was anaerobically mixed with excess 4-(methoxymethyl)phenol in the stopped-flow spectrophotometer at 25°C, pH 7.5 ***(30)***. Spectral scans are shown from 5.6 ms to 965.6 ms with intervals of 80 ms. The absorbance at 450 nm decreases due to flavin reduction. The absorbance increase at 364 nm is ascribed to the formation of the quinone-methide of 4-(methoxymethyl)phenol. The enzyme-bound product intermediate is rather stable under anaerobic conditions. Upon oxygen ingress, the flavin is rapidly reoxidized and the final products 4-hydroxybenzaldehyde and methanol are released.

acts according to a ternary complex mechanism in which the reductive half-reaction is rate-limiting in catalysis ***(30)***.

To study the reductive half-reaction, the oxidized enzyme is mixed with various concentrations of the aromatic substrate in the stopped-flow spectrophotometer under anerobic conditions. Reduction of the flavin is conveniently monitored at 439 nm, the absorption maximum of oxidized VAO. Alternatively, due to the relatively low rate, the reductive half-reaction can be studied by diode array detection. As can be seen from **Fig. 2**, the anerobic reaction of VAO with 4-(methoxymethyl)phenol results in flavin reduction and formation of a species with an intense absorption maximum at 364 nm. Based on the spectral properties of reference compounds and additional evidence, this species has been assigned to the *p*-quinone methide of 4-(methoxymethyl)phenol. As the reduced enzyme-product intermediate complex is rather stable under anerobic conditions, its reoxidation can be studied in the stopped-flow apparatus ***(30)***. From mixing the reduced enzyme-*p*-quinone methide complex with various concentrations of oxygen, the second-order rate constant for reoxidation can be determined. Reoxidation of the enzyme-product intermediate complex is only slightly slower than reoxidation of the free enzyme, and clearly not rate-limiting in overall catalysis. Furthermore, reoxidation of the reduced enzyme-*p*-quinone methide complex is accompanied by the formation of 4-hydroxybenzaldehyde, as evidenced by the increase of absorbance at 335 nm. Moreover, by using $H_2{}^{18}O$ and mass spectral analysis, it was confirmed that the carbonylic oxygen atom from the aromatic product is derived from water ***(30)***.

1.3. Flavoprotein Monooxygenases

Flavoprotein monooxygenases are inducible enzymes that catalyze the insertion of one atom of molecular oxygen into the substrate, using NAD(P)H as external electron donor ***(37,38)***. Most flavoprotein monooxygenases contain FAD as a noncovalently bound prosthetic group and originate from soil microorganisms that can grow on a wide variety of aromatic compounds as sole carbon source. The enzymatic oxygenation of an organic compound requires the activation of oxygen. For flavoprotein monooxygenases, this is achieved by the transient stabilization of a covalent peroxyflavin adduct ***(39)***.

Depending on the type of substrate to be oxygenated, flavoprotein monooxygenases can be divided into: (1) aromatic hydroxylases (electrophilic attack at an activated carbon atom of the substrate aromatic ring) and (2) miscellaneous monooxygenases (oxygenation of either nucleophilic or electrophilic substrates) ***(37)***. These subclasses show remarkable differences in their reaction sequence, i.e., kinetic mechanism ***(40)***. Reactions catalyzed by flavoprotein monooxygenases generally involve three substrates and three products:

$$\mathrm{S + NAD(P) + H^+ + O_2 \xrightarrow{\text{E-FAD}} SOH + NAD(P)^+ + H_2O} \tag{6}$$

As usual for flavoenzymes, this multisubstrate reaction can be separated into two half-reactions. In the reductive half-reaction, the flavin is reduced by NAD(P)H and in the oxidative half-reaction, the reduced flavin is reoxidized by molecular oxygen. With the aromatic hydroxylases, reduction of protein-bound FAD by NAD(P)H is strongly stimulated by binding of the substrate, acting as an effector. After $NAD(P)^+$ release, the reduced flavin reacts rapidly with oxygen to form the labile peroxyflavin species that attacks the substrate aromatic ring. However, with other monooxygenases like cyclohexanone monooxygenase ***(41)***, the substrate binds after formation of the (more stable) peroxyflavin and $NAD(P)^+$ remains bound during the oxygenation reaction.

The catalytic mechanism of flavoprotein monooxygenases has been extensively studied by rapid-reaction techniques. Examples include *p*-hydroxybenzoate hydroxylase ***(42,43)***, melilotate hydroxylase ***(44)***, microsomal flavin-containing monooxygenase ***(45)***, cyclohexanone monooxygenase ***(41)***, anthranilate hydroxylase ***(46)***, phenol hydroxylase ***(47)***, 4-hydroxyphenyl-acetate 3-hydroxylase ***(48)***, 2-aminobenzoyl-CoA monooxygenase/reductase ***(49)***, and 2-methyl-3-hydroxypyridine-5-carboxylic acid oxygenase ***(50)***. Based on properties of *p*-hydroxybenzoate hydroxylase from *Pseudomonas fluorescens*, we here describe a general strategy to study the catalytic mechanism of flavoprotein aromatic hydroxylases.

p-Hydroxybenzoate hydroxylase (PHBH, EC 1.14.13.2) is a homodimer of 88 kDa that catalyzes the NADPH-dependent conversion of 4-hydroxybenzoate to 3,4-dihydroxybenzoate ***(51)***. Much understanding about the PHBH structure has emerged ***(52–55)***, but the interaction between the enzyme and NADPH remains to be elucidated ***(56,57)***. PHBH has a narrow substrate specificity, being active with a limited number of 4-hydroxybenzoate derivatives ***(37)***. Potential substrates such as 4-aminobenzoates are not converted because of their poor effector properties ***(51)***. The activity of PHBH can be determined by measuring the rate of oxygen consumption using an oxygen electrode. This polarographic method also allows for a discrimination between true substrates and nonsubstrate effectors. The latter compounds uncouple the hydroxylation reaction from oxygen reduction, resulting in the formation of hydrogen peroxide. The amount of hydrogen peroxide produced can easily be determined by performing the oxygen consumption experiments in the absence and presence of catalase ***(42)***. The activity of PHBH is most conveniently determined by recording the decrease in absorbance of NADPH at 340 nm. Standard assays and steady-state kinetics preferably are performed at pH 8.0 and 25°C, the pH optimum of overall catalysis ***(58)***. However, pH 6.5 and 4°C allow for a better resolution of individual reaction steps ***(42)***. From steady-state and transient kinetic data it was established that under these conditions, and according to the terminology of Cleland ***(59)***, the enzyme obeys a Bi Uni Uni Uni ping-pong mechanism ***(60)***.

Figure 3 shows the reaction cycle of PHBH, as deduced from transient kinetics. To study the reductive half-reaction, the oxidized enzyme-substrate complex is anearobically mixed with various concentrations of NADPH in the stopped-flow spectrophotometer. Reduction of the flavin is conveniently monitored at 450 nm, the "visible" absorption maximum of the enzyme-substrate complex. Furthermore, small absorption changes above 500 nm are indicative for the transient stabilization of charge transfer complexes between flavin and NADP(H) ***(61,62)***. To study the oxidative half-reaction, the anerobic, reduced enzyme-substrate complex is mixed with oxygen in the stopped-flow spectrophotometer ***(42,43)***. As can be deduced from **Fig. 3**, this reaction involves three consecutive steps with two transient flavin intermediates formed. The first intermediate represents the flavin C(4a)-hydroperoxide oxygenating species

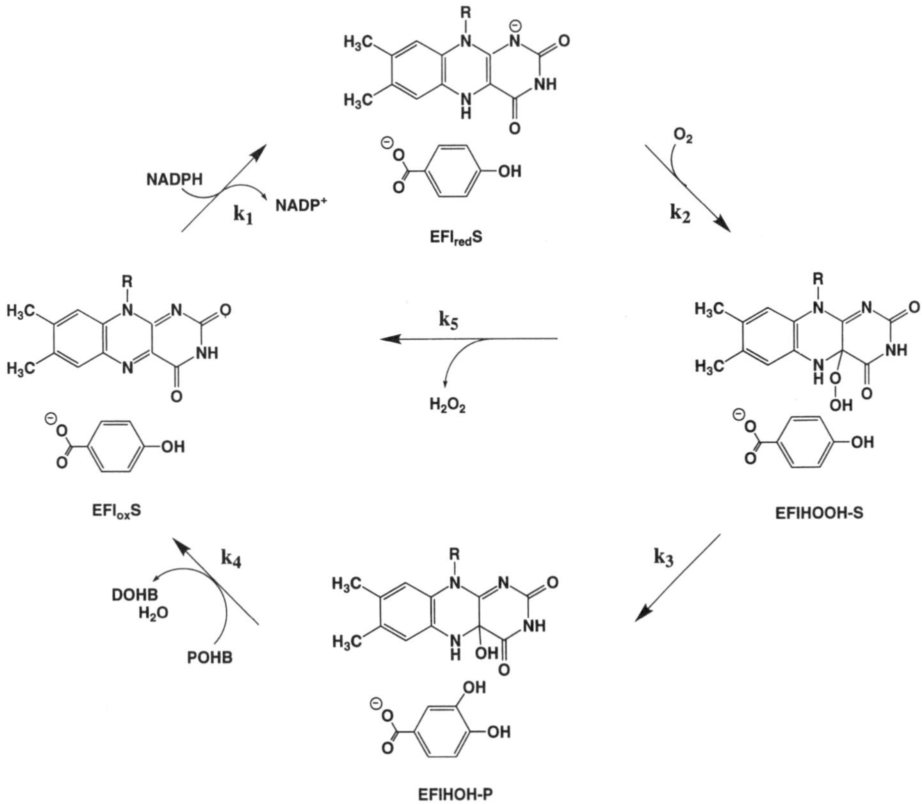

Fig. 3. Reaction cycle of *p*-hydroxybenzoate hydroxylase from *Pseudomonas fluorescens*. Reaction cycle of PHBH as deduced from stopped flow kinetics at 4°C, pH 6.5 ***(42,43,60)***. POHB, 4-hydroxybenzoate; DOHB, 3,4-dihydroxybenzoate; k_1, reduction of enzyme-substrate complex; k_2, formation of flavin C(4a)-hydroperoxide; k_3, substrate hydroxylation; k_4, dehydration of flavin C(4a)-hydroxide and product release; k_5, uncoupling of hydroxylation.

whereas the second intermediate represents the flavin C(4a)-hydroxide, which is formed as a result of substrate hydroxylation. In the final step, this intermediate is dehydrated, regenerating the oxidized enzyme. Because both covalent flavin adducts have similar spectral properties with absorption maxima near 385 nm ***(43)***, the oxidative half-reaction preferably is monitored between 320 and 520 nm at 5 nm intervals. The reductive and oxidative half-reactions can also be studied in the absence of substrate or in the presence of an effector or substrate analog, providing more insight in the ligand-induced stabilization of reaction intermediates ***(51)***. Finally, whenever appropriate, fluorescence and diode array detection are helpful to yield additional information about the occurrence and identity of reaction intermediates ***(47,63)***.

2. Materials

2.1. Lipoamide Dehydrogenase

1. 50 m*M* Sodium pyrophosphate, 0.5 m*M* ethylenediaminetetraacetic acid (EDTA) pH 8.0 (PPi/EDTA buffer).
2. 50 m*M* NAD^+ (Boehringer Mannheim, Mannheim, Germany, grade I) in water. Prepare freshly. Store at 4°C (*see* **Note 1**).
3. 50 m*M* $Lip(SH)_2$ (reduced DL-6,8-thioctic acid amide) in methanol. Prepare freshly. Store at 4°C (*see* **Note 2**).
4. 50 μ*M* LipDH in PPi/EDTA buffer. Store at 4°C. The enzyme concentration is determined from the molar absorption coefficient: ε_{457} = 11.3 mM^{-1} cm^{-1} for protein-bound FAD.
5. Waterbath thermostatted at 25°C.
6. Single-beam spectrophotometer + recorder.
7. Double-beam spectrophotometer thermostatted at 25°C.
8. Stopped-flow spectrophotometer thermostatted at 25°C.
9. Stopped-flow diode array spectrophotometer thermostatted at 25°C.

2.2. Vanillyl-Alcohol Oxidase

1. 50 m*M* Potassium phosphate buffer, pH 7.5 (Pi buffer)
2. 1 m*M* Vanillyl alcohol (4-hydroxy-3-methoxybenzyl alcohol, Aldrich, Milwaukee, WI) in 50 m*M* glycine-NaOH buffer pH 10. Prepare freshly.
3. 2 m*M* 4-(Methoxymethyl)phenol (Aldrich) in Pi buffer. Prepare freshly.
4. 1 *M* Glucose in Pi buffer.
5. 100 μ*M* Glucose oxidase (Boehringer Mannheim, grade I) in Pi buffer. Store at 4°C.
6. 20 μ*M* VAO in Pi buffer. Store at 4°C. The enzyme concentration is determined from the molar absorption coefficient : ε_{439} = 12.5 mM^{-1} cm^{-1} for the covalently bound FAD.
7. 10 m*M* Sodium dithionite in Pi buffer. Prepare freshly.
8. Waterbath thermostatted at 25°C.
9. Single-beam spectrophotometer + recorder.
10. Double-beam spectrophotometer thermostatted at 25°C.
11. Stopped-flow spectrophotometer thermostatted at 25°C.
12. Stopped-flow diode array spectrophotometer thermostatted at 25°C.

2.3. p-Hydroxybenzoate Hydroxylase

1. 100 m*M* Tris/sulfate buffer pH 8.0 (Tris buffer).
2. 50 m*M* Potassium phosphate buffer, pH 6.5 (Pi buffer).
3. 10 m*M* 4-Hydroxybenzoate pH 7.0. Store at 4°C.
4. 10 m*M* NADPH (Boehringer Mannheim, grade I) in Tris buffer. Prepare freshly. Store at 4°C (*see* **Note 3**).

5. 1 *M* Glucose in Pi buffer.
6. 100 μ*M* Glucose oxidase (Boehringer, Mannheim, grade I) in Pi buffer. Store at 4°C.
7. 50 μ*M* PHBH in Pi buffer. Store at 4°C. The enzyme concentration is determined from the molar absorption coefficient of protein-bound FAD: ε_{450} = 10.2 mM^{-1} cm^{-1}.
8. 10 m*M* Sodium dithionite in Tris buffer. Prepare freshly.
9. Waterbath thermostatted at 25°C.
10. Single-beam spectrophotometer + recorder.
11. Double-beam spectrophotometer thermostatted at 25°C.
12. Stopped-flow spectrophotometer thermostatted at 5°C or 25°C.

3. Methods

3.1. Lipoamide Dehydrogenase

3.1.1. Standard Assay

LipDH catalyses the flavin-mediated oxidation of $lip(SH)_2$ by NAD^+. The enzymatic reaction is conveniently monitored by following the increase in absorbance of NADH at 340 nm. Preferably use 1 mL glass cuvets with 1 cm light path.

1. Incubate 950 μL PPi/EDTA buffer at 25°C for 5 min to reach equilibrium at this temperature.
2. Add 20 μL 50 mM NAD^+ and 20 μL 50 m*M* $lip(SH)_2$.
3. Mix carefully and adjust the spectrophotometer to 0.1 absorbance (use an absorbance range 0–1).
4. Start the enzymatic reaction by adding 10 μL 0.5 μ*M* LipDH. Mix carefully and monitor the increase in absorbance at 340 nm during 1 min. Perform the experiment in triplicate.
5. Calculate the specific activity of LipDH in units/mg. The molar absorption coefficient of NADH, ε_{340} = 6.22 mM^{-1} cm^{-1}.

3.1.2. Steady-State Kinetics

For studying steady-state kinetics, the concentration of substrate A (B) is varied at a fixed concentration of B (A). Substrate concentration combinations are chosen such as to avoid errors due to inactivation of the enzyme in the time course of the experiment ***(15)***.

1. Incubate 950 μL PPi/EDTA buffer at 25°C for 5 min to reach equilibrium at this temperature.
2. Add 20 μL of 10 m*M* NAD^+ and 20 μL of 10 m*M* $lip(SH)_2$.
3. Mix carefully and adjust the spectrophotometer to 0.1 absorbance (use an appropriate absorbance range).

4. Start the enzymatic reaction by adding 10 μL 0.5 μ*M* LipDH. Mix carefully and monitor the increase in absorbance at 340 nm during 1 min (*see* **Note 4**).
5. Repeat the experiment with fixed concentrations of NAD^+ and various concentrations of lip$(SH)_2$. Use at least eight substrate concentrations well distributed on the Michaelis-Menten curve and perform each experiment in triplicate.
6. Repeat the experiment with fixed concentrations of lip$(SH)_2$ and various concentrations of NAD^+. Use at least eight substrate concentrations well distributed on the Michaelis-Menten curve and perform each experiment in triplicate.
7. Calculate the apparent enzymatic reaction rate (v_{app} in units/mg) for each assay and construct Lineweaver-Burk plots of the lip$(SH)_2$/NAD^+ reaction with NAD^+ varied and lip$(SH)_2$ fixed and of the lip$(SH)_2$/NAD^+ reaction with lip$(SH)_2$ varied and NAD^+ fixed. The molar absorption coefficient of NADH, $\varepsilon_{340} = 6.22\ mM^{-1}\ cm^{-1}$.
8. Determine the kinetic parameters k_{cat}, K_m lip$(SH)_2$ and K_m NAD^+ from secondary double reciprocal plots.

3.1.3. Rapid-Reaction Kinetics

The reductive and oxidative half-reactions of pig heart, *E. coli,* and *A. vinelandii* LipDH have been studied in detail ***(2,10,12,15,26)***. These studies have revealed that the enzymes share a similar catalytic mechanism but differ with respect to the rate-limiting step and the sensitivity for overreduction to the inactive EH_4 state. The latter property is related to the redox potential of the flavin ***(64)***. The protocols for the reductive and oxidative half-reactions of LipDH are described in this section.

3.1.3.1. Reductive Half-Reaction

By mixing the oxidized enzyme with lip$(SH)_2$, the kinetics of the reductive half-reaction can be studied. The reaction as monitored at 450 nm and 530 nm can essentially be described by a single-exponential function. The reaction includes the first step in the catalytic cycle: binding of the dithiol substrate to form the Michaelis complex $E_{ox}\cdot$lip$(SH)_2$. In the succeeding step the active-site disulfide is reduced by the transfer of two electrons from the substrate, i.e., via thiol interchange. As can be seen from **Eq. 2**, it is assumed that the product lipS_2 rapidly dissociates from EH_2.

1. Prepare at least five solutions with different concentrations (0.2–4.0 m*M*) of lip$(SH)_2$ and a solution of 50 μ*M* LipDH. Since only the L-enantiomer of lip$(SH)_2$ reacts rapidly, substrate concentrations are adjusted to those of the L-enantiomer. The substrate is dissolved in PPi/EDTA buffer/ethanol to yield a final ethanol content of 5% in the mixing chamber.
2. Fill one syringe of the stopped-flow apparatus with the lowest concentration of the substrate solution and the other syringe with the enzyme solution (*see* **Note 5**).

3. Both solutions are mixed in the stopped-flow apparatus and by monitoring the absorbance decrease at 450 nm or the absorbance increase at 530 nm, the apparent rate of reduction (k'_2) of the active-site disulfide can be measured.
4. By changing the substrate concentration, the maximal reduction rate (k_2) and the binding constant ($K_d = k_{-1}/k_1$) can be determined by fitting the data according to the Michaelis-Menten equation ($k'_2 = k_2\,[S]/(K_d + [S])$).

3.1.3.2. Oxidative Half-Reaction

As noted above, the free unliganded EH_2 species can be obtained by careful titration of E_{ox} with sodium borohydride. By mixing the two-electron reduced enzyme with NAD^+ in the stopped-flow spectrophotometer, the kinetics of the oxidative half-reaction can be studied. The reaction as monitored at 450 nm and 700 nm can essentially be described by a single-exponential function. The reaction includes the formation of the Michaelis complex $EH_2{\cdot}NAD^+$ (**Eq. 2**). In the succeeding step, two electrons are transferred from the dithiol via the flavin to NAD^+, resulting in the formation of NADH. Because NAD^+ is used in large excess, the reverse reaction (reduction of E_{ox} by NADH) is catalytically of no significance.

1. Prepare at least five solutions with different concentrations (0.2–4.0 m*M*) of NAD^+ (Boehringer, grade I) and a solution of 50 µ*M* LipDH in PPi buffer (*see* **Note 1**).
2. Prepare EH_2 using sodium borohydride (*see* **Note 6**).
3. Fill one syringe of the stopped-flow apparatus with the lowest concentration of the NAD^+ solution and the other syringe with the enzyme solution.
4. Both solutions are mixed in the stopped-flow apparatus and by monitoring the absorbance decrease at 450 nm or the absorbance increase at 700 nm, the apparent rate of reduction (k'_4) of the flavin (oxidation of the active-site dithiol) can be measured. Apparent rate constants should represent the average of a minimum of four shots.
5. By changing the NAD^+ concentration, the maximal oxidation rate (k_4) and the binding constant ($K_d = k_{-3}/k_3$) can be determined by fitting the data according to the Michaelis-Menten equation ($k'_4 = k_4\,[NAD^+]/(K_d + [NAD^+])$).

3.1.3.3. Reduction of Oxidized Enzyme by NADH

By mixing the oxidized enzyme with NADH in the stopped-flow spectrophotometer, important information can be obtained on the (transient) stabilization of enzyme-pyridine nucleotide complexes and distribution of electronic forms. For that reason, the kinetics of the reduction reaction are monitored at 450 nm, 530 nm and 700 nm or, alternatively, by diode array detection ***(15)***.

Other disulfide reductases like glutathione reductase and thioredoxin reductase use NADPH as the physiological electron donor. For details on the reductive half-reactions of these enzymes, the reader is referred to refs. ***(4)*** and ***(9)***.

1. Prepare at least five solutions with different concentrations (0.2–4.0 m*M*) of NADH (Boehringer, grade I) and a solution of 50 μ*M* LipDH in PPi buffer.
2. Fill one syringe of the stopped-flow apparatus with the lowest concentration of the NADH solution and the other syringe with the enzyme solution (*see* **Note 5**).
3. Mix both solutions in the stopped-flow apparatus and monitor the absorbance decrease at 450 nm or the absorbance increase at 530, respectively 700 nm. Alternatively, follow the reaction by rapid scan analysis, using diode array-detection.
4. Determine rate constants at infinite NADH concentration and K_d NADH values from non-linear fitting of apparent rate constants obtained at, at least five different NADH concentrations. Apparent rate constants should represent the average of a minimum of four shots.

3.2. Vanillyl-Alcohol Oxidase

3.2.1. Standard Assay

VAO catalyses the oxidation of vanillyl alcohol to vanillin. The enzymatic reaction is conveniently monitored by following the increase in absorbance of vanillin at 340 nm. Preferably use 1 mL glass cuvets with 1 cm light path.

1. Incubate 990 μL 1.0 m*M* vanillyl alcohol at 25°C for 5 min to reach equilibrium at this temperature.
2. Adjust the spectrophotometer to 0.1 absorbance (use an absorbance range 0–1).
3. Start the enzymatic reaction by adding 10 μL 20 μ*M* VAO. Mix carefully and monitor the increase in absorbance at 340 nm during 1 min. Perform the experiment in triplicate.
4. Calculate the specific activity of VAO in units/mg. The molar absorption coefficient of vanillin, $\varepsilon_{340\ nm,\ pH\ 10} = 22.8\ mM^{-1}\ cm^{-1}$.

3.2.2. Steady-State Kinetics

The steady-state kinetic parameters for a specific flavoprotein oxidase can be obtained by standard enzyme assays. As VAO obeys Michaelis-Menten kinetics, the K_m values for both substrates (S and O_2) can be determined by varying the concentration of the first substrate at several fixed concentrations of the second substrate. As a result, the maximal turnover rate (k_{cat}) can be determined. Another elegant way to determine these values is by performing so-called enzyme-monitored turnover experiments ***(65)***. By this approach, again the steady-state kinetic parameters can be determined. Furthermore, these type of experiments give an indication whether the rate limiting step is located in the reductive or the oxidative part of the catalytic cycle.

1. Mix 490 μL air-saturated Pi buffer with 500 μL 2 m*M* 4-(methoxymethyl)phenol and incubate at 25°C for 5 min to reach equilibrium at this temperature.
2. Adjust the spectrophotometer to 0.1 absorbance (use an absorbance range 0–1).

3. Start the enzymatic reaction by adding 10 μL 20 μ*M* VAO. Mix carefully and monitor the formation of 4-hydroxybenzaldehyde by the increase in absorbance at 340 nm during 1 min.
4. Repeat the experiment with various concentrations of 4-(methoxymethyl)phenol well distributed on the Michaelis-Menten curve. Perform all experiments in triplicate.
5. Repeat the experiment with fixed concentrations of 4-(methoxymethyl)phenol and various concentrations of oxygen. For this purpose, buffers saturated with 100% nitrogen and 100% oxygen are mixed in different ratios. Perform all experiments in triplicate.
6. Calculate the apparent enzymatic reaction rate (v_{app} in units/mg or k'_{cat} in s^{-1}) for each assay. The molar absorption coefficient of 4-hydroxybenzaldehyde, $\varepsilon_{340\ nm}$, pH 7.5 = 10 mM^{-1} cm^{-1}.
7. Determine the kinetic parameters k_{cat}, K_m 4-(methoxymethyl)phenol and K_m O_2 from secondary double reciprocal plots.

3.2.3. Rapid-Reaction Kinetics

Rapid-reaction studies of VAO have shown that the kinetic mechanism can vary dependent on the type of substrate ***(30,31)***. With most substrates studied, a ternary complex mechanism is operative with flavin reduction being rate-limiting in catalysis.

3.2.3.1. Reductive Half-Reaction

$$E_{ox} + S \underset{k_{-1}}{\overset{k_1}{\rightleftharpoons}} E_{ox}S \underset{k_{-2}}{\overset{k_2}{\rightleftharpoons}} E_{red}P \xrightarrow{k_3} E_{red} + P \qquad (7)$$

By anerobically mixing the oxidized enzyme with substrate, the reductive half-reaction can be studied. The reaction includes the first step in the catalytic cycle: binding of substrate to form the Michaelis-Menten complex ($E_{ox}S$). In the succeeding step the flavin cofactor is reduced by the transfer of two electrons from the substrate. Depending on the substrate, the formed product or product intermediate may already be released from the active-site (k_3) without reoxidation of the reduced flavin. In case of a ternary complex mechanism, this step will be of no significance.

1. To ensure working under anerobic conditions all necessary compartments of the stopped-flow apparatus must be equilibrated with oxygen-free solutions. For this, the chamber containing the syringes can be flushed with oxygen-free argon whereas the syringes, the cell, and all connecting parts should be rinsed with anerobic solutions containing glucose (10 m*M*) and glucose oxidase (0.5 μ*M*).
2. Prepare several solutions with different concentrations (0.1–2.0 m*M*) of substrate and a solution of 20 μ*M* VAO (flasks with airtight septum) which all include glucose (10 m*M*).

3. For anaerobiosis, flush the solutions by oxygen-free argon and subsequently add anerobically, using an airtight syringe, (argon flushed) glucose oxidase (0.2 µ*M*) to all flasks to ensure anerobic conditions (*see* **Note 5**).
4. Fill one syringe of the stopped-flow apparatus with the anerobic substrate solution and the other syringe with the anerobic enzyme solution.
5. Upon mixing both solutions in the stopped-flow apparatus and by monitoring the absorbance at 439 nm, the reduction of the flavin cofactor can be measured. Apparent rate constants should represent the average of a minimum of four shots. Alternatively, with diode-array detection, absorption spectra can be recorded in millisecond time scale. By monitoring spectral changes using diode-array detection, additional information e.g., product formation and/or dissociation can be obtained in a single experiment (*see* **Note 7**; **Fig. 2**).
6. By changing the substrate concentration, the maximal reduction rate (k_2) and the binding constant for the aromatic substrate ($K_d = k_{-1}/k_1$) can be determined by fitting the data according to the Michaelis-Menten equation ($k'_2 = k_2\,[S]/(K_d + [S])$) (*see* **Note 8**).

3.2.3.2. Oxidative Half-Reaction

$$E_{red}\,(L) + O_2 \xrightarrow{k_4} E_{ox}(L) + H_2O_2 \xrightarrow{k_5} E_{ox} + L) \qquad (8)$$

(L represents substrate, product or substrate analog)

The oxidative half-reaction can be examined by mixing the reduced enzyme with molecular oxygen in the stopped-flow spectrophotometer. Reduced flavoprotein oxidases generally react rapidly with molecular oxygen yielding oxidized enzyme and hydrogen peroxide (k_4). Contrary to flavin-dependent monooxygenases, no oxygenated flavin intermediates have ever been detected during this relatively fast process ***(39)***. The rate of reoxidation of reduced flavoprotein oxidases can significantly be influenced by the binding of ligands (product, substrate). This is of importance to test when studying the kinetic mechanism ***(29)***. In practice, measurement of the oxidative half-reaction is similar to the method described for the reductive half-reaction. However, in this case only the syringe containing the reduced enzyme should be anerobic whereas the other syringe should contain an aerated or oxygenated solution. By measuring the increase of flavin absorbance around 450 nm, the rate of reoxidation can be determined. With most flavoproteins oxidases, the reduced enzyme reacts with molecular oxygen without forming a Michaelis-Menten type complex. Therefore, by varying the oxygen concentration and making linear plots of the primary data, one can determine the bimolecular rate constant for this second order process.

1. Prepare a 10-µ*M* solution of reduced VAO by titration of argon flushed enzyme with sodium dithionite (*see* **Note 9**).

2. Fill one syringe of the stopped-flow apparatus with the anerobic reduced enzyme solution and the other syringe with aerated Pi buffer.
3. Upon mixing both solutions in the stopped-flow apparatus and by monitoring the increase in absorbance at 439 nm, the reoxidation of the reduced flavin cofactor can be measured.
4. Repeat the experiment with various concentrations of molecular oxygen. For this purpose, buffers saturated with 100% nitrogen and 100% oxygen are mixed in different ratios. Apparent rate constants should represent the average of a minimum of four shots.
5. Determine the bimolecular rate constant of the reaction of free reduced VAO with molecular oxygen from plotting the rate of reoxidation versus the oxygen concentration.
6. For generation of the reduced enzyme-product intermediate complex, mix 10 μ*M* VAO anerobically with a 1.5 molar excess of 4-(methoxymethyl)phenol.
7. Repeat the reoxidation experiment with the enzyme-product intermediate complex. Apparent rate constants should represent the average of a minimum of four shots.
8. Determine the bimolecular rate constant of the reoxidation of reduced VAO complexed with the quinone methide of 4-(methoxymethyl)phenol from plotting the rate of reoxidation versus the oxygen concentration.

3.3. p-Hydroxybenzoate Hydroxylase

3.3.1. Standard Assay

PHBH catalyzes the NADPH-dependent hydroxylation of 4-hydroxybenzoate to 3,4-dihydroxybenzoate. The enzymatic reaction is conveniently monitored by following the decrease in absorbance of NADPH at 340 nm. Preferably use 1 mL glass cuvets with 1 cm light path.

1. Incubate 950 μL air-saturated Tris buffer at 25°C for 5 min to reach equilibrium at this temperature.
2. Add 20 μL of 10 m*M* 4-hydroxybenzoate and 20 μL of 10 m*M* NADPH.
3. Mix carefully and adjust the spectrophotometer to 0.9 absorbance (use an absorbance range 0–1).
4. Start the enzymatic reaction by adding 10 μL of 2 μ*M* PHBH. Mix carefully and monitor the decrease in absorbance at 340 nm during 1 min. Perform the experiment in triplicate.
5. Calculate the specific activity of PHBH in units/mg. The molar absorption coefficient of NADPH, $\varepsilon_{340\ nm} = 6.22\ mM^{-1}\ cm^{-1}$.

3.3.2. Steady-State Kinetics

The steady-state kinetic parameters for a specific flavoprotein aromatic hydroxylase can be obtained by initial rate measurements of enzyme activity. As PHBH obeys Michaelis-Menten kinetics, the K_m values for all three substrates can be determined by varying the concentration of the first substrate at a constant concentration of the second substrate and a series of fixed concentrations of the third substrate. As a result, the maximal turnover rate (k_{cat}) at infinite concentrations of all the three substrates can be determined ***(60)***.

1. Incubate 950 μL air-saturated Pi buffer at 25°C for 5 min to reach equilibrium at this temperature.
2. Add 20 μL of 5 m*M* 4-hydroxybenzoate and 20 μL of 5 m*M* NADPH.
3. Mix carefully and adjust the spectrophotometer to the desired starting absorbance (use an appropriate absorbance range).
4. Start the enzymatic reaction by adding 10 μL of 2 μ*M* PHBH. Mix carefully and monitor the decrease in absorbance at 340 nm during 1 min. Perform each experiment in triplicate.
5. Repeat the experiment with various concentrations of 4-hydroxybenzoate and several fixed concentrations of NADPH (*see* **Note 10**).
6. Repeat the experiment with various concentrations of 4-hydroxybenzoate at a constant concentration of NADPH and several fixed concentrations of oxygen. For this purpose, buffers saturated with 100% nitrogen and 100% oxygen are mixed in different ratios.
7. Repeat the experiment with various concentrations of NADPH at a constant concentration of 4-hydroxybenzoate and several fixed concentrations of oxygen. For this purpose, buffers saturated with 100% nitrogen and 100% oxygen are mixed in different ratios.
8. Calculate the apparent enzymatic reaction rate (k'_{cat} in s^{-1}) for each assay and construct Lineweaver Burk plots with: (a) 4-hydroxybenzoate varied and oxygen constant at several fixed concentrations of NADPH, (b) 4-hydroxybenzoate varied and NADPH constant at several fixed concentrations of oxygen and (c) NADPH varied and 4-hydroxybenzoate constant at several fixed concentrations of oxygen. The molar absorption coefficient of NADPH, $\varepsilon_{340\ nm} = 6.22\ mM^{-1}\ cm^{-1}$ (*see* **Note 11**).
9. Determine the kinetic parameters k_{cat}, K_m 4-hydroxybenzoate, K_m NADPH and K_m O_2 from secondary double reciprocal plots ***(60)*** .

3.3.3. Rapid-Reaction Kinetics

Rapid-reaction kinetics of flavoprotein aromatic hydroxylases provide insight in the effector specificity (stimulation by aromatic ligands of rate of flavin reduction by NADPH) and in the mechanism and efficiency of substrate hydroxylation (stabilization of oxygenated flavin intermediates).

3.3.3.1. Reductive Half-Reaction

$$E_{ox}S + NADPH \underset{}{\overset{k1}{\rightleftharpoons}} E_{ox}S{:}NADPH \xrightarrow{k2} E_{red}S{:}NADP^{+} \xrightarrow{k3} E_{red}S + NADP^{+} \quad (9)$$

By anerobically mixing the oxidized enzyme-substrate complex with NADPH, the reductive half-reaction of PHBH can be studied. The reaction initially involves ternary complex formation, i.e., binding of NADPH. In the succeeding step, the flavin cofactor is reduced and finally, the $NADP^+$ product is released, leaving the reduced enzyme-substrate complex.

1. To ensure working under anerobic conditions all necessary compartments of the stopped-flow apparatus must be equilibrated with oxygen-free solutions. For this, the chamber containing the syringes can be flushed with oxygen-free argon while the syringes, the cell, and all connecting parts should be rinsed with anerobic solutions containing glucose (10 m*M*) and glucose oxidase (0.5 μ*M*).
2. Prepare several Tris buffer solutions with different concentrations (0.1–2.0 m*M*) of NADPH and a solution of 30 μ*M* PHBH (flasks with airtight septum) which all include 1 m*M* 4-hydroxybenzoate and 10 m*M* glucose.
3. For anaerobiosis, flush the solutions by oxygen-free argon and subsequently add anerobically, using an airtight syringe, (argon flushed) glucose oxidase (0.2 μ*M*) to all flasks to ensure anerobic conditions.
4. Fill one syringe of the stopped-flow appartus with the anerobic enzyme solution and the other syringe with the anerobic NADPH solution.
5. Upon mixing both solutions in the stopped-flow apparatus and by monitoring the absorbance decrease at 450 nm, the reduction of the flavin cofactor can be measured. At temperatures below 10°C, the formation of charge-transfer absorption can be monitored at 690 nm.
6. By varying the NADPH concentration, the maximal reduction rate (k_2) and the binding constant for NADPH ($K_d = k_{-1}/k_1$) can be determined from fitting the traces recorded at 450 nm according to the Michaelis-Menten equation ($k'_2 = k_2$ [NADPH]/(K_d + [NADPH])). Apparent rate constants should represent the average of a minimum of four shots (*see* **Note 12**).

3.3.3.2. Oxidative Half-Reaction

The oxidative half-reaction of PHBH (cf. **Fig. 3**) can be examined by mixing the reduced enzyme-substrate complex with molecular oxygen in the stopped-flow spectrophotometer.

1. To ensure working under anerobic conditions all necessary compartments of the stopped-flow apparatus must be equilibrated with oxygen-free solutions (*see* **Subheading 3.3.3.1.**).
2. For anaerobiosis, flush a solution of 30 μ*M* PHBH in Pi buffer, containing 1 m*M* 4-hydroxybenzoate (flasks with airtight septum), with oxygen-free argon (*see* **Note 5**).
3. Reduce the anerobic enzyme solution by careful titration with a solution of sodium dithionite, using a gas-tight syringe (*see* **Note 9**).
4. Fill one syringe of the stopped-flow appartus with the anerobic enzyme-substrate solution and the other syringe with air-saturated Pi buffer, containing 1 m*M* 4-hydroxybenzoate (*see* **Note 13**).

5. Upon mixing both solutions in the stopped-flow apparatus and by monitoring absorbance changes between 320 and 520 nm, the oxidative half-reaction can be measured (*see* **Note 14**).
6. From varying the concentration of oxygen, the bimolecular rate constant for the reaction of reduced PHBH with molecular oxygen can be determined ***(43,44)***. For this purpose, buffers saturated with 100% nitrogen and 100% oxygen are mixed in different ratios.

4. Notes

1. NAD^+ solutions are relatively stable at pH 6 but rather unstable at pH 8. Prepare dilutions in PPi buffer, immediately prior to use.
2. DL-Dihydrolipoamide is synthesized according to Reed ***(66)***. Dihydrolipoamide solutions are standardized by titration into 5,5'-dithiobis(2-nitrobenzoate) for measurement of free thiol groups ***(64)***.
3. NADPH solutions are relatively stable at pH 8 but rather unstable at pH 6.5. Prepare dilutions in Pi buffer, immediately prior to use.
4. LipDH from *A. vinelandii* is sensitive to dead-end inhibition by NADH ***(19)***. Therefore, for accurate estimation of steady-state kinetic parameters, initial rate measurements of LipDH should preferably be performed using the stopped-flow spectrophotometer. This allows initial velocity measurements from the very beginning of the reaction, with good reproducibility even at very low substrate concentrations ***(14,20)***.
5. Reduction experiments are performed under anerobic conditions. For this, anerobic solutions are made in flasks (or cuvets) with air-tight septum. The flasks are alternatively evacuated and gassed with argon. Before use, the argon is passed over a catalyst (BASF, R-3-11) to remove oxygen followed by passage through an illuminated solution containing methyl viologen, EDTA and proflavin to remove the last traces of oxygen ***(67)***.
6. Free EH_2 can best be prepared in a tonometer ***(64)***. The enzyme solution is contained in the main part of the tonometer, while a few μL of the borohydride solution (5- to 10-fold molar excess) should be located in the sidearm. The sodium borohydride solution is prepared in 10 m*M* NaOH to avoid decomposition and should be as concentrated as possible. The solution in the tonometer is made anerobic by repeated evacuation and flushing with oxygen free argon. When anaerobiosis is established the enzyme solution can be mixed with the sodium borohydride (now dry) in the side-arm. The enzyme will immediately turn red and can be transferred anerobically to one syringe of the stopped-flow spectrophotometer. The substrate solution in the other syringe should be anerobic as well.
7. Rapid reaction kinetics on VAO have been carried out in a Hi-Tech SF-51 stopped-flow apparatus with a dead-time of 1.3 ms (Hi-Tech Scientific, Salisbury, UK). Spectral scans were recorded with a Hi-Tech M300 monochromator diode array-detector with a dead-time of about 6 ms. Spectral scans were collected each 10 ms. Deconvolution analysis of spectral data was performed using the Specfit global analysis program version 2.10 (Spectrum Software Assn. Chapel Hill, NC).

8. With some VAO substrates, it has been found that the reduction reaction is reversible ($k_{-2} > 0$) which results in an incomplete reduction and an apparent finite value for the rate of reduction at infinite low substrate concentration ***(31)***. In that case, the kinetic data can be fitted using a modified Michaelis-Menten equation ***(68)*** or using the fraction of oxidized enzyme formed at the end of the reduction reaction ***(31)***.
9. Sodium dithionite solutions are unstable below pH 7.5 ***(67)***. Stock solutions should be prepared freshly and kept between pH 7.5–8.0.
10. For accurate estimation of steady-state kinetic parameters, initial rate measurements of PHBH should preferably be performed using the stopped-flow spectrophotometer. This allows initial velocity measurements from the very beginning of the reaction, with good reproducibility even at very low substrate concentrations ***(60)***.
11. For accurate estimation of steady-state kinetic parameters, primary data are treated by nonlinear regression analysis.
12. By using the computer simulation program KINSIM ***(69)*** and treating the data according to the scheme depicted in **Eq. 9**, rate constants for hydride transfer and $NADP^+$ dissociation can be generated from best fits of stopped-flow traces recorded at 690 nm. Furthermore, these simulations allow to calculate the contribution of $E_{ox}S$:NADPH and $E_{red}S:NADP^+$ to the charge transfer absorption ***(62)***.
13. For rapid equilibrium, ensure that both syringe solutions contain 1 m*M* 4-hydroxybenzoate.
14. For analysis of rapid reaction data of the oxidative half-reaction and calculation of intermediate spectra, the reader is referred to refs. ***(43,44,63)***.

References

1. Ghisla, S. and Massey, V. (1989) Mechanisms of flavoprotein-catalyzed reactions. *Eur. J. Biochem.* **181,** 1–17.
2. Williams, C. H. Jr. (1992) Lipoamide dehydrogenase, glutathione reductase, thioredoxin reductase, and mercuric ion reductase—a family of flavoenzyme transhydrogenases. In *Chemistry and Biochemistry of Flavoenzymes,* vol. 3 (Müller, F., ed.), CRC Press, Boca Raton, FL, pp. 121–211.
3. Pai, E. F. (1991) Variations on a theme: the family of FAD-dependent NAD(P)H-disulfide-oxidoreductases. *Curr. Opin. Struct. Biol.* **1,** 796–803.
4. Rietveld, P., Arscott, L. D., Berry, A., Scrutton, N. S., Deonarain, M. P., Perham, R. N., and Williams, C. H. Jr. (1994) Reductive and oxidative half-reactions of glutathione reductase from *Escherichia coli. Biochemistry* **33,** 13,888–13,895.
5. Sahlman, L., Lambeir, A. M., Lindskog, S., and Dunford, H. B. (1984) The reaction between NADPH and mercuric reductase from *Pseudomonas aeruginosa. J. Biol. Chem.* **259,** 12,403–12,408.
6. Miller, S. M., Moore, M. J., Massey, V., Williams, C. H. Jr., Distefano, M. D., Ballou, D. P., and Walsh, C. T. (1989) Evidence for the participation of Cys558 and Cys559 at the active site of mercuric reductase. *Biochemistry* **28,** 1194–1205.
7. Williams, C. H. Jr. (1995) Mechanism and structure of thioredoxin reductase from *Escherichia coli. FASEB J.* **9,** 1267–1276.

8. Lennon, B. W. and Williams, C. H. Jr. (1995) Effect of pyridine nucleotide on the oxidative half-reaction of *Escherichia coli* thioredoxin reductase. *Biochemistry* **34,** 3670–3677.
9. Lennon, B. W. and Williams, C. H. Jr. (1997) Reductive half-reaction of thioredoxin reductase from *Escherichia coli. Biochemistry* **36,** 9464–9477.
10. Massey, V., Gibson, Q. H., and Veeger, C. (1960) Intermediates in the catalytic action of lipoyl dehydrogenase. *Biochem. J.* **77,** 341–351.
11. Veeger, C. and Massey, V. (1963) A new NAD^+-dependent spectral species of lipoamide dehydrogenase. *Biochim. Biophys. Acta* **67,** 679–681.
12. Matthews, R. G., Ballou, D. P., and Williams, C. H. Jr. (1979) Reactions of pig heart lipoamide dehydrogenase with pyridine nucleotides—evidence for an effector role for bound oxidized pyridine nucleotide. *J. Biol. Chem.* **254,** 4974–4981.
13. Wilkinson, K. D. and Williams, C. H. Jr. (1981) NADH inhibition and NAD^+ activation of *Escherichia coli* lipoamide dehydrogenase catalyzing the NADH-lipoamide reaction. *J. Biol. Chem.* **256,** 2307–2314.
14. Sahlman, L. and Williams, C. H. Jr. (1989) Lipoamide dehydrogenase from *Escherichia coli*—steady-state kinetics of the physiological reaction. *J. Biol. Chem.* **264,** 8039–8045.
15. Benen, J. A. E., van Berkel, W. J. H., Dieteren, N., Arscott, D., Williams, C. H. Jr., Veeger, C., and de Kok, A. (1992) Lipoamide dehydrogenase from *Azotobacter vinelandii*: site-directed mutagenesis of the His450–Glu455 diad—kinetics of wild-type and mutated enzymes. *Eur. J. Biochem.* **207,** 487–497.
16. Mattevi, A., de Kok, A., and Perham, R. N. (1992) The pyruvate dehydrogenase multienzyme complex. *Curr. Opin. Struct. Biol.* **2,** 877–887.
17. De Kok, A. and van Berkel, W. J. H. (1996) Lipoamide dehdyrogenase, in *Alpha-Keto Acid Dehydrogenase Complexes* (Patel, M. S., Roche, T. E., and Harris, R. A., eds.), Birkhäuser Verlag, Basel, Switzerland, pp. 53–70
18. van Berkel, W. J. H., Benen, J. A. E., and Snoek, M. C. (1991) On the FAD-induced dimerization of apo-lipoamide dehydrogenase from *Azotobacter vinelandii* and *Pseudomonas fluorescens. Eur. J. Biochem.* **197,** 769–779.
19. van Berkel, W. J. H., Regelink, A. G., Beintema, J. J., and de Kok, A. (1991) The conformational stability of the redox states of lipoamide dehydrogenase from *Azotobacter vinelandii. Eur. J. Biochem.* **202,** 1049–1055.
20. Westphal, A. H., Fabisz-Kijowska, A., Kester, H., Obels, P. P., and de Kok, A. (1995) The interaction between lipoamide dehydrogenase and the peripheral-component-binding domain from the *Azotobacter vinelandii* pyruvate dehydrogenase complex. *Eur. J. Biochem.* **234,** 861–870.
21. Massey, V. (1960) The identity of diaphorase and lipoyl dehydrogenase. *Biochim. Biophys. Acta.* **37,** 314–322.
22. Visser, J. and Veeger, C. (1968) Relations between conformations and activities of lipoamide dehydrogenase—protein association-dissociation and the influence on catalytic properties. *Biochim. Biophys. Acta.* **159,** 265–275.
23. Massey, V. and Veeger, C. (1961) Studies on the reaction mechanism of lipoyl dehydrogenase. *Biochim. Biophys. Acta* **48,** 33–47.

24. Benen, J. A. E., van Berkel, W. J. H., Zak, Z., Visser, A. J. W. G., Veeger, C., and de Kok, A. (1991) Lipoamide dehydrogenase from *Azotobacter vinelandii*: site-directed mutagenesis of the His450-Glu455 diad—spectral properties of wild type and mutated enzymes. *Eur. J. Biochem.* **202,** 863–872.
25. Benen, J. A. E., van Berkel, W. J. H., Veeger, C., and de Kok, A. (1992) Lipoamide dehydrogenase from *Azotobacter vinelandii*—the role of the C-terminus in catalysis and dimer stabilization. *Eur. J. Biochem.* **207,** 499–505.
26. Williams, C. H. Jr. (1976) Flavin-containing dehydrogenases, in *The Enzymes,* vol. 13 (Boyer, P. D., ed.), Academic Press, New York, pp. 89–173.
27. Pollegioni, L., Blodig, W., and Ghisla, S. (1997) On the mechanism of D-amino acid oxidase. *J. Biol. Chem.* **272,** 4924–4934.
28. Maeda-Yorita K., Aki, K., Sagai, H., Misaki, H., and Massey, V. (1995) L-lactase oxidase and L-lactate monooxygenase: mechanistic variations on a common structural theme. *Biochimie* **77,** 631–642.
29. Ramsay, R. R. (1991) Kinetic mechanism of monoamine oxidase A. *Biochemistry* **30,** 4624–4629.
30. Fraaije, M. W. and van Berkel, W. J. H. (1997) Catalytic mechanism of the oxidative demethylation of 4-(methoxymethyl)phenol by vanillyl-alcohol oxidase. *J. Biol. Chem.* **272,** 18,111–18,116.
31. Fraaije, M. W., van den Heuvel, R. H. H., Roelofs, J. C. A. A., and van Berkel, W. J. H. (1998) Kinetic mechanism of vanillyl-alcohol oxidase with short-chain 4-alkylphenols. *Eur. J. Biochem.* **253,** 712–719.
32. De Jong, E., van Berkel, W. J. H., van der Zwan, R. P., and de Bont, J. A. M. (1992) Purification and characterization of vanillyl-alcohol oxidase from *Penicillium simplicissimum. Eur. J. Biochem.* **208,** 651–657.
33. Fraaije, M. W., Veeger, C., and van Berkel, W. J. H. (1995) Substrate specificity of flavin-dependent vanillyl-alcohol oxidase from *Penicillium simplicissimum. Eur. J. Biochem.* **234,** 271–277.
34. Fraaije, M. W., Pikkemaat, M., and van Berkel, W. J. H. (1997) Enigmatic gratuitous induction of the covalent flavoprotein vanillyl-alcohol oxidase in *Penicillium simplicissimum. Appl. Environ. Microbiol.* **63,** 435–439.
35. Mattevi, A., Fraaije, M. W., Mozzarelli, A., Olivi, L., Coda, A., and van Berkel, W. J. H. (1997) Crystal structures and inhibitor binding in the octameric flavoenzyme vanillyl-alcohol oxidase: the shape of the active-site cavity controls substrate specificity. *Structure* **5,** 907–920.
36. Fraaije, M. W., van Berkel, W. J. H., Benen, J. A. E., Visser, J., and Mattevi, A. (1998) A novel oxidoreductase family sharing a conserved FAD-binding domain. *Trends Biochem. Sci.* **23,** 206–207.
37. van Berkel, W. J. H. and Müller, F. (1991) Flavin-dependent monooxygenases with special reference to *p*-hydroxybenzoate hydroxylase, in *Chemistry and biochemistry of flavoenzymes,* vol. 2 (Müller, F., ed.), CRC Press, Boca Raton, FL, pp. 1–29.
38. van Berkel, W. J. H., Eppink, M. H. M., van der Bolt, F. J. T., Vervoort, J., Rietjens, I. M. C. M., and Schreuder, H. A. (1997) *p*-Hydroxybenzoate hydroxy-

lase: mutants and mechanism, in *Flavins and Flavoproteins XII* (Stevenson, K., Massey, V. and Williams, C. H. Jr., eds.), University of Calgary Press, Calgary, Alberta, Canada, pp. 305–314.
39. Massey, V. (1994) Activation of molecular oxygen by flavins and flavoproteins. *J. Biol. Chem.* **269,** 22,459–22,462.
40. Müller, F. (1985) Flavin-dependent hydroxylases. *Biochem. Soc. Trans.* **13,** 443–447.
41. Ryerson, C. C., Ballou, D. P., and Walsh, C. (1982) Mechanistic studies on cyclohexanone oxygenase. *Biochemistry* **21,** 2644–2655.
42. Entsch, B., Ballou, D. P., and Massey, V. (1976) Flavin-oxygen derivatives involved in the hydroxylation of *p*-hydroxybenzoate hydroxylase. *J. Biol. Chem.* **251,** 2550–2563.
43. Entsch, B. and Ballou, D. P. (1989) Purification, properties and oxygen reactivity of *p*-hydroxybenzoate hydroxylase from *Pseudomonas aeruginosa*. *Biochim. Biophys. Acta* **999,** 313–322.
44. Schopfer, L. M. and Massey, V. (1980) Kinetic and mechanistic studies of the oxidation of the melilotate hydroxylase-2-hydroxycinnamate complex by molecular oxygen. *J. Biol. Chem.* **255,** 5355–5363.
45. Beaty, N. B. and Ballou, D. P. (1981) The reductive half-reaction of liver microsomal FAD-containing monooxygenase. *J. Biol. Chem.* **256,** 4611–4618.
46. Powlowski, J., Ballou, D. P., and Massey, V. (1989) A rapid reaction study of anthranilate hydroxylase. *J. Biol. Chem.* **264,** 16,008–16,016.
47. Maeda-Yorita, K. and Massey, V. (1993) On the reaction mechanism of phenol hydroxylase. New information obtained by correlation of fluorescence and absorbance stopped flow studies. *J. Biol. Chem.* **268,** 4134–4144.
48. Aranuchalam, U. and Massey, V. (1994) Studies on the oxidative half-reaction of *p*-hydroxyphenylacetate 3-hydroxylase. *J. Biol. Chem.* **269,** 11,795–11,801.
49. Langkau, B. and Ghisla, S. (1995) Kinetic and mechanistic studies on the reactions of 2-aminobenzoyl-CoA monooxygenase/reductase. *Eur. J. Biochem.* **230,** 686–697.
50. Chaiyen, P., Brisette, P., Ballou, D. P., and Massey, V. (1997) Unusual mechanism of oxygen atom transfer and product rearrangement in the catalytic reaction of 2-methyl-3-hydroxypyridine-5-carboxylic acid oxygenase. *Biochemistry* **36,** 8060–8070.
51. Entsch, B. and van Berkel, W. J. H. (1995) Structure and mechanism of *para*-hydroxybenzoate hydroxylase. *FASEB J.* **9,** 476–483.
52. Schreuder, H. A., Prick, P., Wierenga, R. K., Vriend, G., Wilson, K. S., Hol, W. G. J., and Drenth, J. (1989) Crystal structure of the *p*-hydroxybenzoate hydroxylase-substrate complex refined at 1.9 Å resolution. *J. Mol. Biol.* **208,** 679–696.
53. Gatti, D. L., Palfey, B. A., Lah, M. S., Entsch, B., Massey, V., Ballou, D. P., and Ludwig, M. L. (1994) The mobile flavin of 4-OH benzoate hydroxylase. *Science* **266,** 110–114.
54. Schreuder, H. A., Mattevi, A., Obmolova, G., Kalk, K. H., Hol, W. G. J., van der Bolt, F. J. T., and van Berkel, W. J. H. (1994) Crystal structures of wild-type *p*-hydroxybenzoate hydroxylase complexed with 4-aminobenzoate, 2,4-dihydroxybenzoate and 2-hydroxy-4-aminobenzoate and the Tyr222Ala mutant, complexed with 2-hydroxy-4-aminobenzoate. Evidence for a proton channel and a new binding mode of the flavin ring. *Biochemistry* **33,** 10,161–10,170.

55. van Berkel, W. J. H., Eppink, M. H. M., and Schreuder, H. A. (1994) Crystal structure of *p*-hydroxybenzoate hydroxylase reconstituted with the modified FAD present in alcohol oxidase from methylotrophic yeasts: Evidence for an arabino-flavin. *Protein Sci.* **3,** 2245–2253.
56. Eppink, M. H. M., Schreuder, H. A., and van Berkel, W. J. H. (1997) Identification of a novel conserved sequence motif in flavoprotein hydroxylases with a putative dual function in FAD/NAD(P)H binding. *Protein Sci.* **6,** 2454–2458.
57. Eppink, M. H. M., Schreuder, H. A., and van Berkel, W. J. H. (1998) Interdomain binding of NADPH in *p*-hydroxybenzoate hydroxylase as suggested by kinetic, crystallographic and modeling studies of His162 and Arg269 variants. *J. Biol. Chem.* **273,** 21,031–21,039.
58. van Berkel, W. J. H. and Müller, F. (1989) The temperature and pH dependence of some properties of *p*-hydroxybenzoate hydroxylase from *Pseudomonas fluorescens*. *Eur. J. Biochem.* **179,** 307–314.
59. Cleland, W. W. (1970) Steady-state kinetics, in *The Enzymes,* vol. 2 (Boyer, P. D., ed.), Academic Press, New York, pp.15–65.
60. Husain, M. and Massey, V. (1979) Kinetic studies on the reaction mechanism of *p*-hydroxybenzoate hydroxylase. Agreement of steady state and rapid reaction data. *J. Biol. Chem.* **254,** 6657–6666.
61. Howell, L. G., Spector, T., and Massey, V. (1972) Purification and properties of *p*-hydroxybenzoate hydroxylase from *Pseudomonas fluorescens*. *J. Biol. Chem.* **247,** 4340–4350.
62. Eppink, M. H. M., Schreuder, H. A., and van Berkel, W. J. H. (1995) Structure and function of mutant Arg44Lys of 4-hydroxybenzoate hydroxylase. Implications for NADPH binding. *Eur. J. Biochem.* **231,** 157–165.
63. Palfey, B. A., Ballou, D. P., and Massey, V. (1997) Flavin conformational changes in the catalytic cycle of *p*-hydroxybenzoate hydroxylase substituted with 6-azido- and 6-aminoflavin adenine dinucleotide. *Biochemistry* **36,** 15,713–15,723.
64. Matthews, R. G. and Williams, C. H. Jr. (1976) Measurement of the oxidation-reduction potentials for two-electron and four-electron reduction of lipoamide dehydrogenase from pig heart. *J. Biol. Chem.* **251,** 3956–3964.
65. Gibson, Q. H., Swoboda, B. E. P., and Massey, V. (1964) Kinetics and mechanism of glucose oxidase. *J. Biol. Chem.* **239,** 3927–3934.
66. Reed, J. K. (1973) Studies on the kinetic mechanism of lipoamide dehydrogenase from rat liver mitochondria. *J. Biol. Chem.* **248,** 4834–4839.
67. Mayhew, S. G. (1978) The redox potential of dithionite and SO_2^- from equilibrium reactions with flavodoxins, methyl viologen and hydrogen plus hydrogenase. *Eur. J. Biochem.* **85,** 535–547.
68. Strickland, S., Palmer, G., and Massey, V. (1975) Determination of dissociation constants and specific rate constants of enzyme-substrate (or protein-ligand) interactions from rapid reaction kinetic data. *J. Biol. Chem.* **250,** 4048–4052.
69. Barshop, B. A., Wrenn, R. F., and Frieden, C. (1983) Analysis of numerical methods for computer simulation of kinetic processes: development of KINSIM, a flexible portable system. *Anal. Biochem.* **130,** 134–145.

6

Application of Freeze Quenching to the Study of Rapid Reactions in Flavoproteins

W. John Ingledew

1. Introduction

Freeze quench entails arresting a reaction by rapid freezing and sustained low temperature, after a specified reaction time. The trapped species are then analyzed. This technique was introduced in a 1961 paper entitled "Sudden freezing as a technique for the study of rapid reactions" by Bray *(1)*, associated papers describe the application of this technique to electron paramagnetic resonance (EPR) *(2)* and studies on xanthine oxidase using these methods *(3)*. The study of enzyme reactions utilizing this technique was first reviewed about 35 yr ago *(4–6)* and the methods are still as used then, although the apparatus has improved.

Study of the intermediates formed during presteady-state kinetics of enzyme catalyzed reactions is extensively used to obtain information about the mechanisms of these enzymes (*see* Chapter 5). The most useful forms of spectroscopy for such investigations can be applied to the analyses in real time and these will always be the preferred techniques. Occasionally, however, the information required can only be obtained by the application of spectroscopies which cannot be applied in real time, e.g., some applications of EPR. For these spectroscopies the reaction has to be stopped (quenched) at known lapsed times (t) after initiation and the spectroscopic indicators assessed "at leisure" (as a snap shot at time *t*). The most common quench technique is chemical quench where the reaction mixture is sprayed into acid (for example) to stop the reaction, after which the products can be analyzed. In freeze quenching the reaction mixture is sprayed into a cold bath (approx –140°C) where reactions are stopped by freezing and low-temperature. Freeze quench has been used in combination with: EPR *(6)* and related spectroscopies *(7)*, Resonance Raman *(8)*

From: Methods in Molecular Biology, vol. 131: Flavoprotein Protocols
Edited by: S. K. Chapman and G. A. Reid © Humana Press Inc., Totowa, NJ

and EXAFS spectroscopy (extended X-ray absorption fine structure) ***(9,10)***. These techniques, when applied to enzymes, cannot always be used in real time because the spectra have to be obtained at low temperatures, or the data collection is slow, or the spectrometer sensitivity at higher temperatures is too low.

The technique of freeze quench has been validated and most used for studies of the milk flavoenzyme xanthine oxidase, mainly in combination with EPR. As early as 1961, Bray established Mo(V) as a transient intermediate redox state during enzyme turnover ***(3)***. Xanthine oxidase contains a molybdenum, a flavin adenine dinucleotide (FAD) and two iron-sulphur clusters. The mechanism by which this enzyme hydroxylates purines has been extensively studied and a comprehensive reactions scheme derived ***(11)***. EPR studies using freeze quench show that the Mo center, flavin and both iron-sulphur clusters undergo redox changes in a time scale compatible with a role in enzyme turnover. Radical forms of Mo (Mo(V)), flavin semiquinone and the iron-sulphur clusters are detected. The xanthine (or analogue) is bound at the Mo center and is rapidly oxidized, reducing the Mo from the (VI) to the (IV) state. The (IV)-state is short lived as electrons are redistributed through the redox centers and the EPR detectable Mo(V) state peaks in under 50 ms then declines. The enzyme is reoxidized by O_2 giving peroxide.

The flavocytochrome *b2* has also been studied by freeze quench with EPR sectroscopy. This enzyme couples L-lactate oxidation to cytochrome *c* reduction and its catalytic cycle has been described in terms of five consecutive electron transfer events ***(12)***. The first reaction is the two-electron reduction of flavin mononucleotide (FMN) (yielding pyruvate). The electrons are then sequentially passed to b_2-haem (intramolecular transfer) and then on to cytochrome *c* (intermolecular transfer). Daff et al. ***(13)*** have investigated this reaction using EPR and freeze quench. Under the reaction conditions a steady-state is reached in 50 ms, at which time approx 75% of the FMN is in the semiquinone state. Subsequently, the levels of the oxidant (cytochrome *c*) are depleted and the free radical concentration falls off rapidly. These observations were entirely consistent with, and gave evidence in support of, the proposed mechanism derived from optical studies ***(12)***.

Freeze quench has also recently been used in the analysis of turnover in nonflavin enzymes. *E. coli* ribonucleotide reductase catalyzes the production of deoxyribonucleotides from ribonucleotides and is composed of two nonidentical subunits. Subunit 1 contains the substrate binding domain and five redox active cysteines and subunit 2 contains a diiron cluster and a tyrosine residue which forms a tyrosyl radical during turnover. Rapid freeze quench coupled to ENDOR (electron-nuclear double resonance) spectroscopy has been used to characterize this radical ***(7)*** and EXAFS with freeze quench has been used to

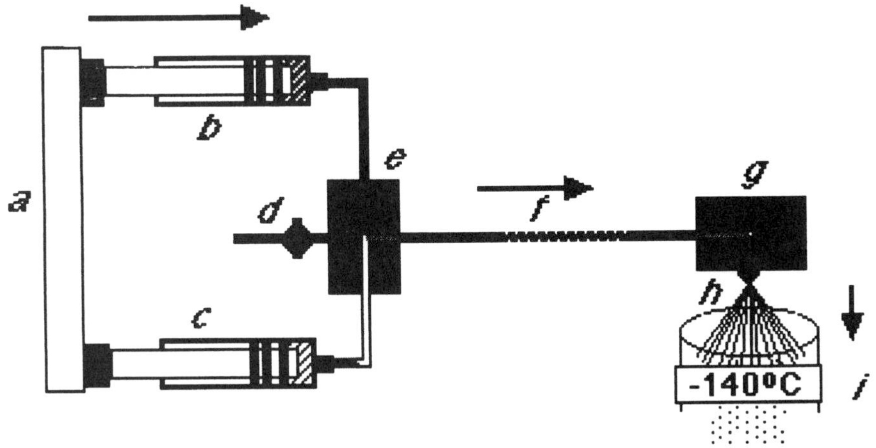

Fig. 1. A schematic representation of the mixing part of a freeze quench apparatus. (a) Ram block, this is motor driven and computer controlled, (b and c) syringes containing the reactants, (d) valve to enable the system to be flushed, (e) mixer, (f) reaction tube, (g) spray nozzle, (h) jet, (i) flute containing quenching solvent at –140°C.

characterize the diiron-tyrosyl radical complex ***(10)***. Other recent applications of freeze quench to kinetic analyses include: a resonance Raman spectroscopic study of the carbon monoxide dehydrogenase from *Clostridium thermoaceticum* ***(8)***; EPR studies to establish the kinetic competence of a radical intermediate in the reaction mechanism in lysine 2,3-aminomutase ***(14)***; a study of the copper in dopamine β-monooxygenase to establish the role of the copper in catalysis ***(15)***, and a study of *E. coli* cytochrome *bo* (a quinol oxidase) seeking evidence for a Q-cycle ***(16)***.

2. Materials

1. The basic freeze quench apparatus: Our freeze quench apparatus is similar to others but with an improved cryostat design ***(17,18)***. The fundamental parts of a freeze quench apparatus are:
   a. a mixing plus aging system,
   b. a quenching system.

   These are represented in the simplified diagram, **Fig. 1**. The mixing/aging system consists of a computer controlled ram *(a)* that actuates two (or more) syringes *(b* and *c)*. The syringe outputs are mixed in a simple T-mixer (*e*) and the mixed reactants travel down a specified length of tube (*f*) to age before being sprayed (*g*,*h*) into the quenching apparatus (*i*). In practice, the reaction tubes are coiled as the distance between *e* and *g* is fixed. The prefreezing reaction age of the sample could be varied either by changing the flow rate or the length of tubing (*f*) along which the reactant mixture travels. Normally we use only two different flow rates with a range of different tube lengths giving overlapping reactions time ranges (different diameter tubes are also used).

2. The quenching occurs when the reaction is stopped by freezing the reactants in precooled isopentane (or similar solvent) at approx –140°C. This is achieved by spraying (*g* and *h*) into the cold isopentane (*i*). Isopentane is the most commonly used solvent for freezing as it is liquid at room temperature (bp 28°C), has a low freezing point (–160°C) and experience has shown it does not appear to denature the reactants ***(6)***. Propane has also been used but is more difficult and hazardous to use.
3. The resolution limit for our apparatus combines the mixing time and the freezing time and is approx 5 ms, which is similar to other systems reported ***(19)***. The efficiency of our mixing system was characterized by chemical quench after the method of Barman and Gutfreund ***(20)*** and of the complete mix-freeze quench system by the reaction of azide with met-myoglobin, after Ballou and Palmer ***(20)***.

3. Methods

1. The reactants are loaded into the syringes through three-way stainless steel taps (not shown). We use glass syringes fitted with triple O-ringed stainless steel plungers (*b* and *c*).
2. Control of our ram system is by a programmed stepping motor control unit ***(17)*** after Gutteridge ***(21)***. On actuation the contents of the syringes are mixed; the mixing and flow devices are stainless steel high-performance liquid chromatography (HPLC) parts (Anachem Ltd., Bedfordshire, UK). The use of stainless steel HPLC parts facilitates maintenance of anaerobic conditions, which has often been a problem in this type of experiment and appears to abolish stretching problems which can be experienced with plastic tubing. HPLC tubing of internal diameters 0.0254–0.0762 cm (cat. nos. U-105, U-106 and U-107 (Anachem Ltd.)) are used with a 0.0254-cm diameter spray head (half of capillary union junction (cat. no. U-412)). We use flow rates of 1 mL/s^{-1} for timings between 20 ms and 235 ms and 0.74 mL/s^{-1} for the narrower bore tubing used for below 10 ms. A simple T-mixer, an HPLC "zero dead volume," four-way coupling cross with the fourth exit closed during mixing (*e*) is employed, this gives, at 1 mL/s^{-1} an exit velocity of 2 m/s^{-1}. The fourth port on the mixer is controlled by a two-way valve (*d*) and used to purge the mixer between experiments (with water and then N_2). On initiation the reactants are forced through the system and sprayed into the quenching apparatus.
3. Temperature control of the reaction, if not performed at room temperature, is achieved by situating the mixing system in a perspex vessel which can function as a temperature regulated "water bath" if required.
4. A simplified representation of the quenching apparatus is given in **Fig. 2**. The quenching vessel is a glass flute (approx vol 60 mL) with an EPR tube at the bottom (the vessel can be one piece or the EPR tube part can be detachable) ***(2,19)***. The flute sits in a cradle and is place inside the cryostat. Temperature control of the quenching vessel is through liquid nitrogen boil off through heating circuits and feedback loops as published ***(17,18)***. The flute is cooled on all

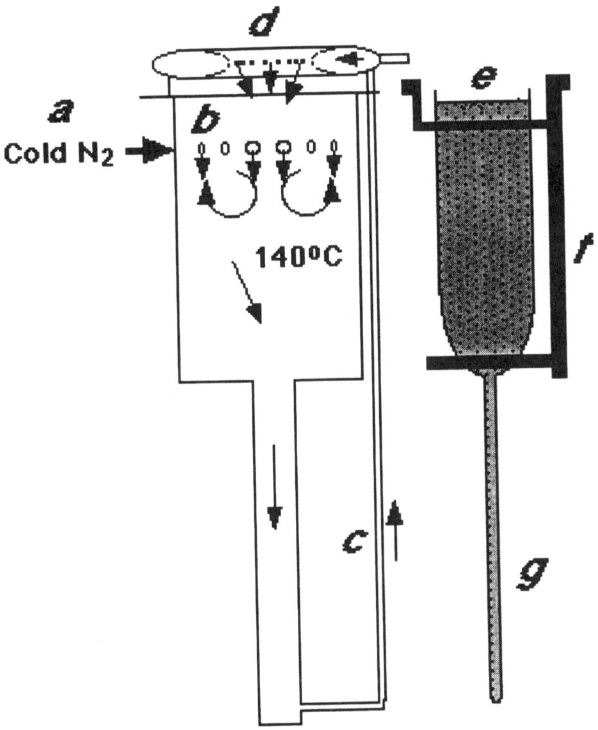

Fig. 2. A schematic representation of the quenching apparatus. The flute (e) is filled with isopentane and sits in a holder (f). The holder is placed in the cryostat chamber (b) (the two are shown separately in the diagram). Cold nitrogen gas (from liquid N2 boil-off and heating) enters the chamber from a ring (a) and is forced down around the EPR tube end of the flute when the flute and holder are in place (g) and evacuates through a portal ring (d via c) across the surface of the isopentane. The temperature of the isopentane is also monitored by a probe.

sides by cold gas flowing around it. The cold gas is from liquid N_2 boil off driven by a heater and brought to the required temperature by secondary heaters. The cryostat is constructed such that it sits in and plugs a liquid N_2 container. Earlier cryostats were hazardous, using large open baths of solvent. This redesigned vessel gives a more convenient and safer apparatus. The air gap between the spray nozzle and the quench fluid is kept "oxygen free" by nitrogen gassing over the solvent surface from a blanketing ring (d). The isopentane is sparged with N_2 gas from a cylinder during cool down to remove oxygen and mix the solvent.

5. After mixing, aging and spraying into cold solvent the reaction mix is in the form of a semicrystalline powder (*see* refs. ***2*** and ***6*** for further details and a photograph of this process). This settles in the cold solvent and is then packed down into the bottom of the 3 mm I.D. EPR tubes using a fine plunger, the head of which snugly

fits the EPR tube but has grooves in it to allow the solvent to pass, *(2)*. The packing is awkward but important. The EPR spectrometer only monitors the bottom 1.5 cm of the tube but a consistent packing density in this region of the tube is essential for reproducible results. If a nonreacting standard is incorporated into the mix it can be used to correct for packing density. This is not often possible and packing is a crucial part of the method and is a major source of experimental noise. If internal standards are added to one of the syringes they must possess an EPR spectrum that does not interfere spectrally with reactants and products and they should not react with the experimental reactants. The packing is done *in situ* in the cryostat. Excess solvent is then aspirated off and the flute removed and stored under liquid N_2 until required. If a two-section flute is used the EPR tube part is detached and stored. Time between trapping and measuring should be kept to a minimum *(6)*.

4. Notes

1. Stopping the reaction. The underpinning rationale of this technique depends on stopping the reaction by freezing and low temperature. Can stopping the reaction be relied on? Often it can, but this is a potential problem and the results must be compared to and be consistent with those obtained by other spectroscopies. The technique depends on slowing the reaction to the extent that there is no significant change between the freeze quench event and the spectrometry. This may include the period during storage under liquid N_2. In this technique trapping is at –140°C and the sample remains at that temperature for approx 2 min followed by storage under liquid N_2 or spectrometry at low temperature.
2. Temperature dependence of electron transfer reactions. Electron transfer theory implicates free energy, distance, and temperature terms: the impact of the first two terms is clearly visualized empirically *(22)*. Temperature dependence is more complex. As different electron transfer mechanisms or different rate limiting processes may dominate at different temperatures the overall process may not have the same temperature dependence throughout the temperature range. Reactions may be very temperature sensitive at higher temperatures (exponentially related to reciprocal of temperature) but may become temperature insensitive at lower temperatures as the temperature sensitive process "freeze" out and tunneling dominates. This is apparent in the study of DeVault and Chance *(23)* who reported that the electron transfer rate between a cytochrome *c* and a light activated bacteriochlorophyll occurred with a $t_{1/2}$ at room temperature of 2 μs, decreasing by three orders of magnitude to 2.3 ms at 100 K but thereafter was relatively temperature independent to 4 K. Substrate turnover at a catalytic site (as opposed to simple electron transfer) is normally very temperature sensitive and reaction intermediates have been readily trapped and stabilized by low-temperature techniques. Low temperature triple trapping studies by Chance and colleagues on respiratory systems show temperature dependence of a number electron transfer reactions, although some of these will be intermolecular reactions or catalytic cycle reactions *(24,25)*.

Although freezing and low temperature are related. They can be considered separately in this context.

3. Freezing. Freezing the reactants stops mobility and thereby stops reactions involving collision of reactants. Freezing may also be expected to impede reactions that require gross protein conformational changes such as folding of a hinge region to bring reacting groups closer together. As a rule of thumb, these types of reaction will be stopped by freezing.
4. Low temperature. See **Note 2** for for caveats concerning the temperature dependence of electron transfer reactions. The low temperature trapping protocols of Chance and colleagues *(24,25)* show an exponential rate dependence on the reciprocal of temperature for formation of compound B in cytochrome *c* oxidase, within the range studied (to approx –120°C). This compound (and related compounds) have been shown to be stable for long periods under liquid N_2, as extrapolation of their data below their experimental range (to –140°C and 77 K) would predict.

   As a general rule intramolecular electron transfer will usually continue at finite rates at low temperatures, equilibrating to a thermodynamic minimum. But parts of the catalytic cycle can be expected to be effectively frozen and intermolecular reactions should stop.
5. The importance of the choice of buffers. When solutions freeze, pH changes occur. These must be minimized by choice of the best buffers for these purposes. This topic is reviewed in refs. ***26*** and ***27***.

References

1. Bray, R. C. (1961) Sudden freezing as a technique for the study of rapid reactions. *Biochem. J.* **81,** 189–193.
2. Bray, R. C. and Petterson, R. (1961) Electron-spin-resonance measurements. *Biochem. J.* **81,** 194–195.
3. Bray, R. C. (1961) The chemistry of xanthine oxidase. 8. Electron-spin-resonance measurements during the enzymic reaction. *Biochem. J.* **81,** 196–199.
4. Bray, R. C. (1964) Quenching by squirting into cold immiscible liquids, in R*apid Mixing and Sampling Techniques in Biochemistry* (Chance, B., Eisenhardt, R. H., Gibson, A. H., and Lonberg-Holm, K. K., eds.), Academic, New York, pp. 195–203.
5. Palmer, G. and Beinert, H. (1964) An experimental evaluation of the Bray rapid freezing technique, in *Rapid Mixing and Sampling Techniques in Biochemistry* (Chance, B., Eisenhardt, R. H., Gibson, A. H., and Lonberg-Holm, K. K., eds.), Academic, New York, pp. 205–218.
6. Bray, R. C. and George, G. N. (1985) Electron paramagnetic-resonance studies using presteady-state kinetics with substitution with stable isotope on the mechanism of action of molybdoenzymes. *Biochem. Soc. Trans.* **13,** 560–567.
7. Burdi, D., Sturgeon, B. E., Wong, H. T., Stubbe, J., and Hoffman, B. M. (1996) Rapid freeze-quench ENDOR of the radical X intermediate of *Escherichia coli* ribonucleotide reductase using $^{17}O_2$, $H_2{}^{17}O$ and ^2H_2O. *J. Am. Chem. Soc.* **118,** 281–282.

8. Qiu, D., Kumar, M., Ragsdale, S. W., and Spiro, T. G. (1995) Freeze-quench resonance Raman spectroscopic evidence for an Fe-CO adduct during acetyl-CoA synthesis and Ni involvement in CO oxidation by carbon monoxide dehydrogenase from *Clostridium thermoaceticum. J. Am. Chem. Soc.* **117,** 2653–2654.
9. George, G. N., Bray, R. C., and Cramer, S. P. (1986) Extended X-ray absorption fine structure studies of transient species during xanthine oxidase turnover by using rapid freezing. *Biochem. Soc. Trans.* **14,** 651–652.
10. Riggs-Gelasco, P. J., Shu, L., Chen, S, Burdi, D., Huynh, B. H., Que, L., and Stubbe, J. (1998) EXAFS characterisation of the intermediate X generated during the assembly of the *Escherichia coli* ribonucleotide reductase R2 diferric tyrosyl radical cofactor. *J. Am. Chem. Soc.* **120,** 849–860.
11. Edmondson, D., Ballou, D., Van Heuvelen, A., Palmer, G., and Massey, V. (1973) Kinetic studies on the substrate reduction of xanthine oxidase. *J. Biol. Chem.* **248,** 6135–6144.
12. Miles, C. S., Rouviere-Fourmy, N., Lederer, F., Mathews, F. S., Reid, G. A., and Chapman, S. F. (1992) Tyr-143 facilitates interdomain electron transfer in flavocytochrome *b2*. *Biochem. J.* **285,** 187–192.
13. Daff, S. Ingledew, W. J., Reid, G. A., and Chapman, S. K. (1996) New insights into the catalytic cycle of flavocytochrome *b2*. *Biochemistry* **35,** 6345–6350.
14. Chang, C. H., Ballinger, M. D., Reed, G. H., and Frey, P. A. (1996) Lysine 2,3-aminomutase: rapid mix-freeze-quench electron paramagnetic resonance studies establishing the kinetic competence of a substrate-based radical intermediate. *Biochemistry* **35,** 11,081–11,084.
15. Brenner, M. C., Murray, C. J., and Klinman, J. P. (1989) Rapid freeze- and chemical-quench studies of dopamine β-monooxygenase: comparison of pre-steady state and steady-state parameters. *Biochemistry* **28,** 4656–4664.
16. Schultz, B. E., Edmondson, D. E., and Chan, S. I. (1998) Reaction of *E. coli* cytochrome *bo* with substoichiometric ubiquinol-2: a freeze quench EPR investigation. *Biochemistry* **37,** 4160–4168.
17. Moodie, A. D., Mitchell, R. H., and Ingledew, W. J. (1990) A gas-flow cryostat for use in freeze-quench studies: design and application to discontinuous pre-steady-state spectral analyses. *Analytical Biochemistry* **189,** 103–106.
18. Moodie, A. D. (1990) Characterisation of the catalytic mechanisms of *Escherichia coli* fumarate reductase and ubiquinol oxidase cytochrome *bd*. Ph.D. thesis, University of St. Andrews, Fife, UK.
19. Ballou, D. P. and Palmer, G. A. (1974) Practical rapid quenching instrument for the study of reaction mechanisms by electron paramagnetic resonance spectroscopy. *Anal. Chem.* **46,** 1248–1252.
20. Barman, T. E. and Gutfreund, H. (1964) A comparison of the resolution of chemical and optical sampling, in *Rapid Mixing and Sampling Techniques in Biochemistry* (Chance, B., Eisenhardt, R. H., Gibson, A. H., and Lonberg-Holm, K. K., eds.), Academic, New York, pp. 195–203.
21. Gutteridge, S., Tanner, S. J., and Bray, R. C. (1978) The molybdenum centre in native xanthine oxidase. *Biochem. J.* **175,** 869–878.

22. Moser, C. C., Keske, J. M., Warncke, K., Farid, S., and Dutton, P. L. (1992) Nature of biological electron transfer. *Nature* **355,** 796–802.
23. DeVault D. and Chance, B. (1966) Studies of photosynthesis using a pulsed laser, 1. Temperature dependence of cytochrome oxidation rate in *Chromatium*. Evidence for tunnelling. *Biophys. J.* **6,** 825–847.
24. Chance, B., Graham, N., and Legallais, V. (1975) Low temperature trapping method for Cytochrome oxidase. *Anal. Biochem.* **67,** 552–579.
25. Chance, B., Saronio, C., and Leigh, J. S. (1975) Functional intermediates in the reaction of membrane-bound cytochrome oxidase with oxygen. *J. Biol. Chem.* **250,** 9226–9237.
26. Orii, K. and Morita, M. (1977) Measurement of the pH of frozen buffer solutions by using pH indicators. *J. Biochem.* **81,** 163–168.
27. Williams-Smith, D. L., Bray, R. C., Barber, M. J., Tsopanakir, A. D., and Vincent, S. P. (1977) Changes in apparent pH on freezing aqueous buffer solutions and their relevance to electron paramagnetic resonance spectroscopy. *Biochem. J.* **167,** 593–600.

7

Application of Electron Spin Resonance (ESR) for Detection and Characterization of Flavoprotein Semiquinones

Marat B. Murataliev

1. Introduction

Flavoproteins are involved in a variety of biological electron transfer processes, from energy accumulation in respiratory and photosynthetic electron transfer chains to oxidation of toxic compounds by cytochrome P450-dependent systems. The structure and function of free and protein bound flavins have been a subject of extensive studies. This effort has resulted in our current understanding of the chemistry and significance of flavins in living cells, as summarized in a number of reviews *(1–4)*.

1.1. Properties of Flavin Semiquinones

The most common flavin cofactors found in flavoproteins are flavin mononucleotide (FMN) and flavin adenine dinucleotide (FAD). Unlike metalloproteins and nicotinamide nucleotides which can undergo only one- and two-electron transitions, respectively, the isoalloxazine moiety of flavin cofactors can undergo both two- and one-electron reactions with formation of three redox states (**Fig. 1**): quinone (oxidized), semiquinone (one electron reduced), and hydroquinone (two electron reduced). This is possible because of the ability of flavins to form relatively stable semiquinone and hydroquinone forms, which are necessary intermediates in the transition between two-electron and one-electron donors/acceptors.

The structures and absorption spectra of model flavin semiquinone compounds have been thoroughly characterized *(5–10)*. Flavin semiquinone can exist in cationic, neutral and anionic forms *(9)*. The structures of these species are shown in **Fig. 1**. The p*K* values for formation of cationic and anionic

From: Methods in Molecular Biology, vol. 131: Flavoprotein Protocols
Edited by: S. K. Chapman and G. A. Reid © Humana Press Inc., Totowa, NJ

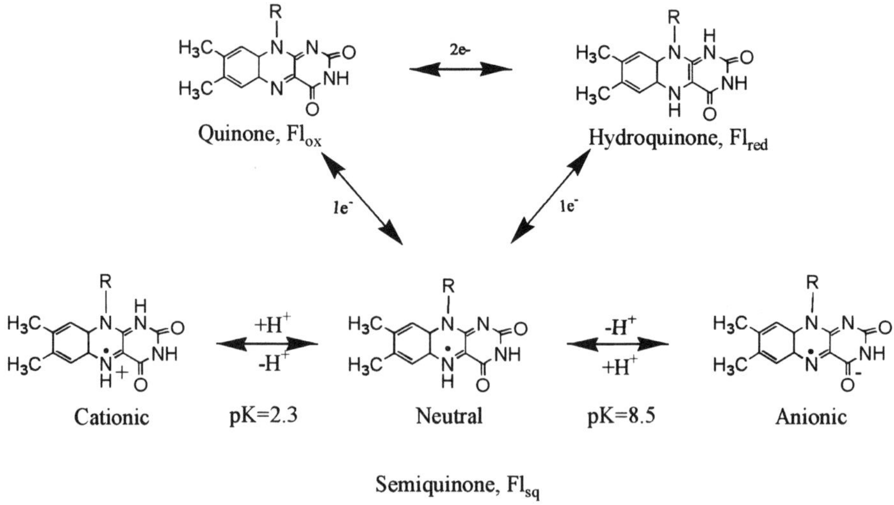

Fig. 1. Structure and redox transitions of flavin.

semiquinone forms for free flavins in solution are 2.3 and 8.5, respectively *(9)*. Because only two semiquinone forms, neutral and anionic, have been found in flavoproteins, properties of the cationic form will not be considered here. Anionic and neutral flavin semiquinones have clearly different light absorption properties ***(10,11)***, *see* Chapter 1. The neutral flavin semiquinone has a characteristic wide absorbance band at about 500–650 nm with a maximum at 580–600 nm. In contrast, the anionic semiquinone has very low absorbance above 550 nm, and a maximum at about 370 nm. Due to the color of their concentrated solutions, neutral and anionic semiquinones are also called blue and red, respectively.

While the overall shape of the spectra of flavin cofactors (in oxidized or semiquinone forms) in flavoproteins does not change much, the positions of the absorbance maxima, as well as their extinction coefficients and peak ratio can vary in different flavoproteins. These variations likely reflect different environments of the flavin binding site, as the spectral properties of free flavins depend on the polarity and pH of the solvent ***(10)***. Binding to a protein also changes redox properties of the flavin cofactors. Redox potentials of the protein-bound flavin cofactors range from –500 mV to +80 mV, as compared to a value of about –200 mV for all redox transitions of free flavins at neutral pH (*see* review, **ref. *2***). The effect of the protein environment on the properties of bound FMN has been demonstrated recently with *Anabaena* flavodoxin. Substitution of a single amino acid residue in the FMN binding site altered redox potentials of one-electron reduction by as much as 137 mV ***(12)***. The mutations also significantly altered absorption and fluorescence spectra of both oxidized and semiquinone forms.

Characterization of the mechanism of electron transfer by flavoproteins necessarily involves detection and identification of the intermediates, particularly flavin semiquinones. The identification and characterization of flavoprotein semiquinones based on their absorption spectra is not reliable in many cases due to the presence of other chromophores such as heme or iron-sulfur clusters, or due to the turbidity of microsomal or mitochondrial suspensions. Formation of the so-called charge-transfer complexes ***(13,14)*** with a wide absorbance band at 500–700 nm can also interfere with spectroscopic detection of the neutral flavin semiquinone. These uncertainties with spectral identification make electron spin resonance (ESR), also called electron paramagnetic resonance (EPR), the only reliable, strong evidence for the formation of a flavin semiquinone.

In this chapter the basic techniques used for detection and characterization of semiquinone forms of flavoproteins by ESR spectroscopy will be presented. The well characterized glucose oxidase is used as a model system. The enzymological significance of the flavin semiquinones in flavoproteins will also be briefly discussed using the results obtained in our group with cytochrome P450BM3. An attempt has been made to emphasize the interpretation of ESR results in relation to the catalytic competence of flavoprotein semiquinone forms.

1.2. Nature of the ESR Signal

The theory and application of ESR can be found in a number of textbooks and reviews ***(15–18)***. Here, an oversimplified description of the ESR phenomenon is presented. The spins of electrons in molecules are usually paired, so that most molecules with an even number of electrons have no net magnetic moment due to the electrons. Such molecules do not respond to a magnetic field in the way required to yield an ESR signal. On the other hand, semiquinones have an odd number of electrons and carry an electron-spin magnetic moment detectable by ESR spectroscopy.

In the absence of an external magnetic field the spin magnetic moments of the unpaired electron are randomly oriented. When an external magnetic field is applied, the electron magnetic moments will become oriented either parallel or antiparallel to the external field. Application of the resonant microwave radiation to the paramagnetic compound placed in the magnetic field results in absorption of the microwave radiation. This absorption will change the orientation of the magnetic moments to the opposite. A fraction of electronic magnetic moments oriented parallel to the field slightly exceeds that of the antiparallel orientation because of the lower energy level. Due to this difference, the change of spin orientation at the resonance condition is associated with a net absorption of energy of the applied electromagnetic radiation, which is detected by a spectrometer to produce an ESR signal.

The interactions of the unpaired electrons with several atomic nuclei split the ESR spectra into a unique pattern, called hyperfine structure. Free semiquinone radicals give rise to spectra exhibiting a number of lines disposed in a centrisymmetrical pattern *(9)*. The hyperfine coupling constants of flavin semiquinones in solution have been determined which allowed estimation of the distribution of spin densities in the isoalloxazine ring *(7,8)*. However, when the flavin coenzyme is bound to the apoprotein, the resolution of the ESR spectrum in aqueous solution completely disappears because of the slow rotational tumbling of the large protein molecule. Thus little insight in the structure of the flavoprotein semiquinone can be obtained, and ESR characterization of the semiquinone forms of flavoproteins is limited to determining basic ESR parameters. These are the g-value and the concentration of the free radical, linewidth and saturation behavior, which allow the ionic state of the flavin semiquinone to be determined.

2. ESR Methodology

2.1. Sample Preparation

Samples are analyzed in quartz tubes 2–5 mm in diameter. For most radicals, spectrometers are able to detect μ*M* concentrations with the sample vol in the range of 100–300 μL. Due to relatively easy saturation of ESR signal of the flavin semiquinones, concentration of up to 100 μ*M* or higher may be required. Another important factor in determining the concentration of a flavoprotein in the ESR sample is the yield of the enzyme containing free radical: a 10 × higher concentration is required when 0.1 mol/mol is formed than when the semiquinone yield is close to 1.0.

Stable semiquinone forms can be analyzed at room temperature. The flavoprotein is incubated under the particular condition leading to the semiquinone formation, the sample is transferred into an ESR tube, and spectra are recorded. However, the semiquinone form of some flavoproteins may be readily oxidized by oxygen, and/or disproportionate at room temperature. Such samples must be frozen in liquid nitrogen and analyzed in the frozen state. ESR spectroscopy of the frozen samples increases the sensitivity of the detection several-fold.

If the flavin semiquinone is readily oxidized by oxygen, the samples for ESR analysis must be prepared and analyzed under anerobic conditions. Oxygen from the solutions is removed by repeated degassing under reduced pressure and flushed with oxygen-free gas, such as nitrogen or argon. The flavoprotein is treated anerobically to induce semiquinone formation. Semiquinone formation can be achieved by a prolonged illumination of the enzyme solution in the presence of ethylenediamine tetraacetic acid (EDTA) (***11***, *see* Chapter 1), or by anerobic addition of a substoichiometric amounts of reducing agent (sodium dithionite, sodium borohydride, NAD(P)H or other).

Another approach to prepare an ESR sample of an unstable flavoprotein semiquinone is the rapid freezing technique described in detail in Chapter 6. Briefly, the solutions to be combined to induce semiquinone formation are rapidly (in ms scale) delivered into a mixing chamber by two air or motor driven syringes. The mixed solution is pushed out of the mixing chamber through the reaction tubing, where it ages. Variation of the length (volume) of the reaction tubing allows different reaction times. For rapid freezing, the reaction mixture is collected into isopentane chilled to –150°C. The ice crystals formed are collected, packed into an ESR tube and analyzed.

2.2. Recording ESR Spectra

The ESR spectra are recorded with an X-band spectrometer operating at a microwave frequency of about 10 GHz, a magnetic field modulation frequency of 100 kHz, and a modulation amplitude of 1–10 G. A modulation amplitude of 1–2 G ($\leq$ 0.1 linewidth) gives a better resolution and more accurate estimation of linewidth, while higher amplitude increases the intensity of the signal. The microwave power settings depend on the experimental goals: better resolution is achieved at nonsaturating microwave power, while the intensity of the signal increases at partial saturation.

3. Methods

3.1. Data Collection

1. A solution of glucose oxidase (Sigma Chemical Co., St. Louis, MO, cat. no. G-7141), 50 μ*M* in flavin, is placed in either spectrophotometer cell or ESR tube and made anerobic by repeated degassing and flushing with oxygen-free argon.
2. The anerobic solution of glucose oxidase is illuminated in the spectrophotometer cell and spectra recorded before and after illumination. For the ESR measurements, illumination is carried out directly in ESR tubes.
3. The samples are placed in a 10°C water bath and illumination is carried out with a white light produced by a Schott KL 1500 microscope illuminator equipped with a 150-W lamp (Schott Corp., Yonkers, NY). The fiberoptic light guides of the illuminator are placed 1/2 in. away from the sample. The ESR samples are frozen in liquid nitrogen after the illumination. A slide projector can be used as the light source for the illumination. It is important to keep the sample in the water bath to avoid overheating and protein denaturation. Time of illumination should be determined experimentally for a particular flavoprotein and light source.

4. ESR spectra are recorded at 77 K with a Bruker model ESP 300E spectrometer (Bruker Instruments, Billerica, MA) operating at microwave frequency of 9.45 GHz, magnetic field modulation frequency of 100 kHz, and modulation amplitude of 1.0 G. The magnetic field is scanned between 3260 and 3460 G. Samples are analyzed in quartz tubes 4 mm in diameter. Two solutions, 10 and 50 μ*M*, of 2,2,6,6-tetramethylpiperidine-*N*-oxyl (TEMPO) in methanol-H_2O (98:2) are used as the standards.

3.2. Results and Interpretation

3.2.1 Formation of the Flavin Semiquinone

The application of ESR spectroscopy to study flavoprotein semiquinones is demonstrated below with glucose oxidase as a model. This FAD containing flavoprotein is known to form easily two types of flavin semiquinone at different pH values, neutral at pH 6.0 and anionic at pH 9.2 ***(11)***. Glucose oxidase solution is illuminated for 150 min and formation of flavin semiquinone is monitored by optical spectroscopy. **Figure 2** shows the absorbance spectra of oxidized glucose oxidase and two semiquinone forms obtained, neutral and anionic. The neutral flavin semiquinone has a characteristic wide absorbance band at about 500–650 nm with a maximum at 580 nm and a shoulder at 630 nm. In contrast, the anionic semiquinone has a very low absorbance above 550 nm, but high absorbance at about 370 nm.

3.2.2. ESR Spectroscopy of Glucose Oxidase Semiquinones

The samples for ESR spectroscopy are prepared by the same procedure, except that photoirradiation is carried out directly in the ESR tubes. After 150 min illumination, the samples are frozen in liquid nitrogen and ESR spectra are recorded. The ESR spectra are usually presented as the first derivative of the absorption line. The spectra of the two flavin semiquinone forms of glucose oxidase are shown in **Fig. 3**.

3.2.3. Linewidth of ESR Spectra

The neutral and anionic flavin semiquinones can be distinguished by different linewidths of the spectra ***(19)***. The linewidth is defined as the separation, in gauss, between the positive and negative extrema in the first derivative presentation, and reflects different hyperfine interaction constants in these two flavin semiquinone species ***(7–9)***. A 19-G linewidth is characteristic for a neutral semiquinone, a 14–15 G linewidth is characteristic for an anionic semiquinone ***(19)***. The linewidth of the spectra of neutral and anionic semiquinone forms of glucose oxidase, shown in **Fig. 3**, are 19.5 and 14.8 G, respectively. The linewidth difference arises from the

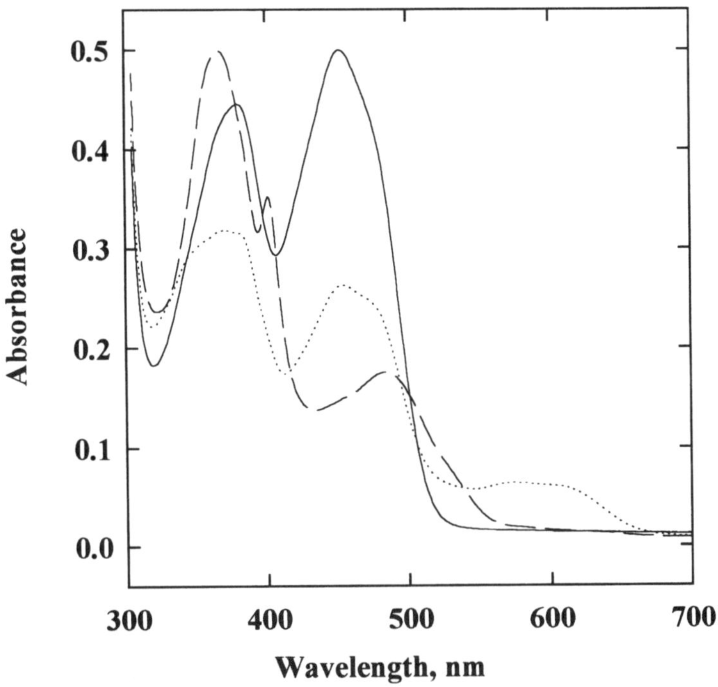

Fig. 2. Absorption spectra of oxidized (solid line), neutral (dotted line) and anionic (dashed line) semiquinone forms of glucose oxidase. The anaerobic enzyme solution (50 μ*M* flavin concentration) was illuminated for 150 min in 150 m*M* Na-phosphate, pH 6.0 or 133 m*M* Tris-HCl, pH 9.2 buffer.

presence of strongly coupled proton at the $N_{(5)}$ position of the neutral semiquinone. Due to the presence of this exchangeable proton, the linewidth of the neutral flavin semiquinone decreases from 19 to about 15 G in D_2O, whereas the linewidth of the anionic semiquinone is unchanged ***(19)***. It should be noted that covalent attachment of the isoalloxazine ring the protein also results in a decrease of the linewidth.

3.2.4. The g-Factor

The *g*-factor qualitatively corresponds to the wavelength maximum in optical spectroscopy, and is an essential but not unique property of the semiquinone. Quantitatively, the *g*-factor characterizes the molecule in which an unpaired electron is located. The *g*-factor for a free electron is 2.0023. The factors close to 2.00 correspond to a highly delocalized molecular orbital of the unpaired electron, while significant deviation suggests strong electron spin interaction with a single atom.

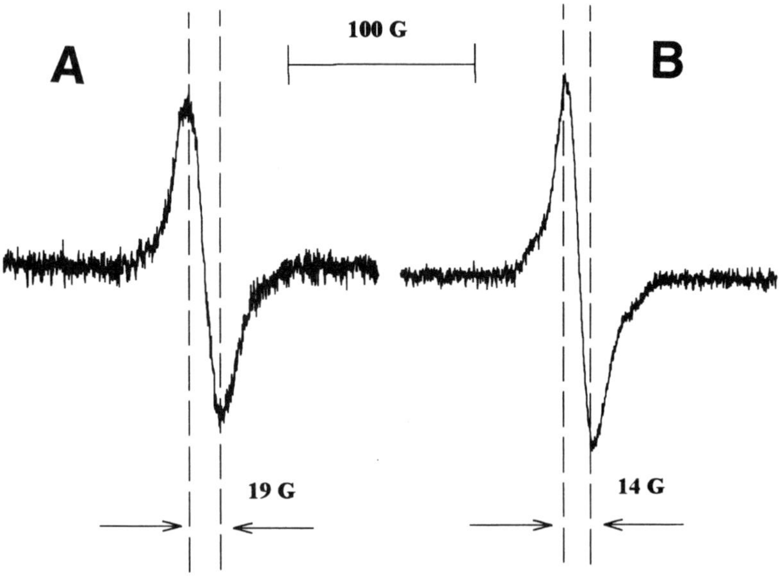

Fig. 3. ESR spectra of the neutral (A) and anionic (B) semiquinone forms of glucose oxidase. Samples were prepared by 150 min photoirradiation of the anaerobic glucose oxidase solution in ESR tubes, and then frozen in liquid nitrogen.

Determination of the *g*-factor requires measurements of the resonant magnetic field (H) and of the microwave frequency (ν). A number of systematic errors are inherent to this approach. Alternatively, the *g*-factors can be determined accurately by comparison of the position of the unknown spectra on the magnetic field scale with that of a stable radical standard with known *g*-factor. The *g*-factor in this case is calculated based on the following equation, where subscripts *s* and *x* stand for standard and unknown samples, respectively, and $\Delta H = H_s - H_x$.

$$g_x = g_s H_s / H_x = g_s H_s / (H_s - \Delta H) = g_s (1 - [\Delta H / H_s])$$

This approach is useful when the unknown and standard samples have close *g*-factors. A number of commercially available compounds with a *g*-factor comparable to that of flavin semiquinones are available as standards: peroxylamine disulfonate and 2,2,6,6-tetramethyl piperidine nitroxides ($g = 2.0054$), and 2,2-diphenyl-1-picrylhydrazine ($g = 2.0036$ in benzene). We used TEMPO as the standard ($g = 2.0054$). For neutral and anionic semiquinone forms of glucose oxidase (**Fig. 3**), *g*-factors are 2.004 and 2.003, respectively. The *g*-factors of the flavin radicals do not depend on the ionization state ***(6)***, in agreement with the fact that the unpaired electron in both forms is largely delocalized.

3.2.5. Quantitation of the Semiquinone Content

The area under the absorption curve of the ESR signal is proportional to the concentration of free radicals. Spectrometer software integrates the signal and determines its area. For accurate quantitation, the area must be measured at a microwave power well below saturation. A stable paramagnetic standard in a crystalline form, or as a standardized stable solution is used for spin quantitation. For better accuracy, the concentration and *g*-factor of the standard compound should be comparable to that of the unknown sample. Standard and analyzed samples should be measured in the same cuvet since the sensitive vol may vary from cuvet to cuvet. Using two or more standard solutions is a good idea to confirm linearity of the scale. Spin concentration is calculated according to the following equation, where C is the concentration, A is the integrated absorption area, subscripts s and x stand for standard and unknown samples, respectively.

$$C_x = C_s A_x / A_s$$

Such an approach allows determination of spin concentration with about ±10% accuracy. Using 10 and 50 μ*M* solutions of TEMPO in methanol-H_2O (98:2) as the standards, the semiquinone content of the glucose oxidase samples is found to be 0.65 and 0.72 mol/mol of flavin for neutral and anionic semiquinone forms, respectively.

3.2.6. Signal Saturation

In addition to a different linewidth, anionic and neutral semiquinones show different saturation behavior as the microwave power is increased ***(19)***. In the absence of saturation, the intensity of the ESR signal is proportional to the microwave power. As a certain value of the microwave power is reached, the dependence deviates from linear and shows saturation at higher microwave power. Spectra of the semiquinone sample are recorded at different microwave power settings, and the signal area is plotted as a function of the microwave power. Saturation curves for the neutral and anionic semiquinone forms of glucose oxidase are shown in **Fig. 4**. The area of the ESR signals of the two semiquinones at lowest microwave power are normalized for direct comparison of the saturation curves, and plotted as a function of the output power of the klystron. Anionic and neutral semiquinones have clearly different saturation behavior.

Half-saturating power reflects a magnitude of relaxation time of flavin semiquinone, and is related to the rate at which energy of the absorbed microwave irradiation can be transferred to the surrounding structure. Even though the measurement of relaxation times by this approach is not very accurate, measuring saturation curves can distinguish different semiquinone forms. Such an

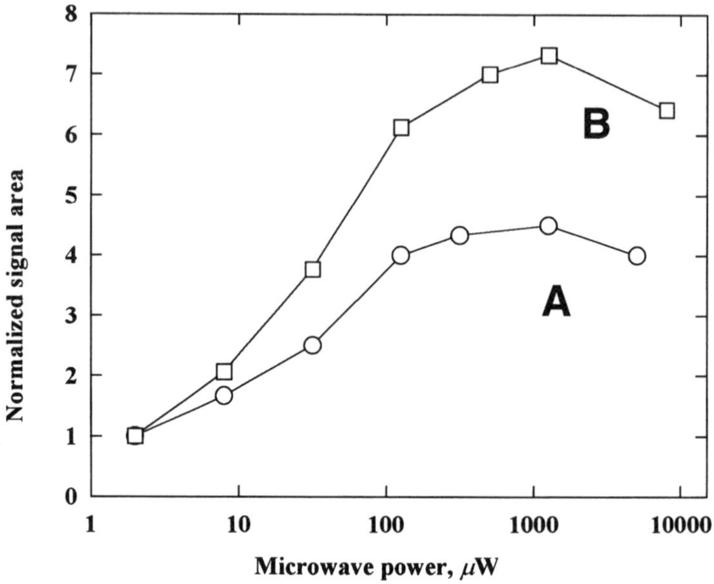

Fig. 4. Saturation curves for the neutral (A) and anionic (B) semiquinone forms of glucose oxidase. Samples were prepared by photoirradiation of the anaerobic glucose oxidase solution in ESR tubes, frozen in liquid nitrogen. ESR spectra were recorded at different output power of the klystron. Nominal power of the klystron at 0 dB attenuation is equal to 200 mW.

approach allowed us to identify and characterize anionic and neutral semiquinone forms of cytochrome P450BM3 ***(20)***.

In some cases, ESR signal of flavin semiquinone shows a phenomenon called anomalous saturation ***(19)***. The room temperature spectra show wings of 70–80 G linewidth, the intensity of which increases with increasing microwave power when the signal of 15 or 19 G is already saturated. Hyde et al. ***(21)*** have discussed the origin of the anomalous saturation behavior. This anomalous saturation is observed at temperatures above –30°C, but disappears at lower temperatures.

The results with glucose oxidase presented in this chapter show how ESR spectroscopy is used for characterization of flavoprotein semiquinones. This approach does not allow more detailed information to be obtained on the structure of the semiquinone or on its environment. Electron double resonance (ELDOR and ENDOR) and electron spin echo techniques can provide information on the structure and environment of the protein bound flavin semiquinone, as well as on spin interaction with surrounding nuclides ***(16,22,23)***.

3.3. ESR Spectroscopy and Enzyme Catalysis

The transitions between different reduction states of free flavins are reversible processes. An equilibrium distribution between the reduction states is governed by the redox potentials of the two one-electron reduction reactions (**Fig. 1**), and a certain concentration of flavin semiquinone can be expected in partially reduced systems. Two important points should be kept in mind when interpreting ESR experiments. The first one concerns the conditions leading to semiquinone formation, and the second one is the significance of the semiquinone for catalysis.

3.3.1. Flavin Semiquinone Formation

In general, almost any flavoprotein will form a flavin semiquinone under specific conditions. Often this is readily achieved by a partial reduction of the flavoprotein. However, the amount and type of the semiquinone formed by flavoproteins can depend on the binding of a substrate, product, or other ligand, as well as on the reaction time. For example, glucono δ-lactone binding to a neutral flavin semiquinone form of glucose oxidase induces formation of the anionic flavin semiquinone *(13)*. Microsomal cytochrome *b5* reductase forms anionic FAD semiquinone only upon binding of NAD^+ *(24)*, whereas mitochondrial adrenodoxin reductase requires presence of a large excess of NADPH to form FAD semiquinone *(25)*. Short-time incubation of P450BM3 with NADPH produces a two-electron reduced enzyme with two types (neutral and anionic) of flavin semiquinone present *(20)*. In contrast, chemical reduction fails to produce significant amounts of semiquinone in a two-electron reduced P450BM3 *(26,27)*.

3.3.2. Catalytic Competence of Flavin Semiquinones

Even though detection of flavin semiquinone formation is an essential step in determining redox properties and the catalytic mechanism of a flavoprotein, in many cases evidence for the catalytic competence of the semiquinone form is required. This is particularly important when nonphysiological conditions, such as anaerobiosis, or light irradiation in the presence of EDTA, or chemical reduction have been used to produce the semiquinone. A number of flavoproteins are known to form a flavin semiquinone which is clearly not a catalytic intermediate. For example, monoamine oxidase B forms two types of flavin semiquinone, only one of which serves as the catalytic intermediate *(28)*. Glucose oxidase and D-amino acid oxidase readily form and stabilize a flavin semiquinone under anerobic conditions, but their catalytic turnover involves two-electron transitions, so that semiquinone forms are unlikely to emerge during catalysis *(13)*. In contrast, cytochrome *b5* reductase and adrenodoxin reductase, whose catalytic cycle involves an FAD semiquinone intermediate, do not form significant amounts of a stable semiquinone upon chemical reduction ***(24,25,29)***. The three-electron reduced diflavin domain of P450BM3 contains virtually stoichiometric amount of semiquinone (likely FAD), but is unable to reduce heme and therefore is catalytically inactive.

Cytochrome P450BM3 is a soluble fatty acid hydroxylase from *Bacillus megaterium* ***(30)***. The enzyme consists of two distinct domains, an FAD and FMN containing reductase and a hemoprotein (CYP102), fused in a single polypeptide ***(30,31)***. Reduced nicotinamide adenine dinucleotide phosphate (NADPH) delivers two reducing equivalents to FAD in the form of a hydride ion. FAD transfers reducing equivalents to FMN, which delivers electrons to the P450 domain. For a catalytic turnover to occur in P450BM3, two electrons need to be delivered to the heme catalytic site in two consecutive one-electron transfer steps. P450BM3 reduction with substoichiometric NADPH rapidly results in formation of a two-electron reduced flavoprotein with a pair of electrons distributed between FAD and FMN ***(20)***. If FMN semiquinone can serve as an electron donor for the heme domain, the enzyme should complete catalytic turnover in the presence of substoichiometric NADPH. Measurements of fatty acid hydroxylation at [NADPH] < [P450BM3] have shown that P450BM3 indeed catalyzes single-turnover fatty acid hydroxylation in the presence of limited NADPH. At a P450BM3 concentration of 10 μM, the addition of 1–9 μM NADPH resulted in rapid laurate hydroxylation with an efficiency of about 80%, thus demonstrating that FMN semiquinone serves as a catalytic intermediate. Similarly, FMN semiquinone serves as an electron donor for cytochrome c.

This is in contrast to mammalian cytochrome P450 reductase, which is homologous to the flavoprotein domain of P450BM3. The mammalian enzyme readily forms an FMN semiquinone, also known as the air-stable semiquinone due to its remarkable stability in the presence of atmospheric oxygen ***(32)***. This form does not reduce cytochrome c despite the fact that redox potential difference favors cytochrome c reduction at equilibrium by a factor of 100 or more.

Acknowledgment

This work was supported by NIH grant GM39014.

References

1. Edmondson, D. E. (1978) ESR of free radicals in enzyme systems, in *Biological Magnetic Resonance*, vol. 1 (Berliner, L. J. and Reuben, J., eds.), Plenum, New York, pp. 205–237.
2. Müller, F. (1983) The flavin redox-system and its biological function. *in: Topics in Current Chemistry*, vol. 108, *Radicals in Biochemistry*, Springer Verlag, New York, pp. 71–107.

3. Edmondson, D. E. and Tollin, G. (1983) Semiquinone formation in flavo- and metalloflavoproteins, in *Topics in Current Chemistry*, vol. 108, Springer Verlag, New York, pp. 109–138.
4. Ghisla, S. and Massey, V. (1989) Mechanisms of flavoprotein-catalyzed reactions. *Eur. J. Biochem.* **181,** 1–17.
5. von Dudley, K. H., Eriksson, L. E. G., and Ehrenberg, A., and Müller, F. (1964) Spektren und strukturen der am flavin-redoxsystem beteiligten partikeln. *Helv. Chim. Acta* **47,** 1354–1383.
6. Eriksson, L. E. G. and Ehrenberg, A. (1964) Electron spin resonance study on anionic flavin free radical. *Acta Chem. Scand.* **18,** 1437–1453.
7. Ehrenberg, A., Müller, F., and Hemmerich, P. (1967) Basicity, visible spectra, and electron spin resonance of flavosemiquinone anions. *Eur. J. Biochem.* **2,** 286–293.
8. Müller, F., Hemmerich, P., Ehrenberg, A., Palmer, G., and Massey, V. (1970) The chemical and electronic structure of the neutral flavin radical as revealed by electron spin resonance spectroscopy of chemically and isotopically substituted derivatives. *Eur. J. Biochem.* **14,** 185–196.
9. Müller, F., Hemmerich, P., and Ehrenberg, A. (1969) On the molecular and submolecular structure of flavin free radicals and their properties, in *Flavin and Flavoproteins* (Kamin, H., ed.), University Park Press, Baltimore, MD, pp. 107–122.
10. Müller, F., Brüstlein, M., Hemmerich, P., Massey, V., and Walker, W. H. (1972) Light-absorption studies on neutral flavin radicals. *Eur. J. Biochem.* **25,** 573–580.
11. Massey, V. and Palmer, G. (1966) On the existence of spectrally distinct classes of flavoprotein semiquinones. A new method for the quantitative production of flavoprotein semiquinones. *Biochemistry* **5,** 3181–3189.
12. Lostao, A., Gomez-Moreno, C., Mayhew, S. G., and Sancho, J. (1997) Differential stabilization of the three FMN redox forms by tyrosine 94 and tryptophan 57 in flavodoxin from *Anabaena* and its influence on the redox potentials. *Biochemistry* **36,** 14,334–13,344.
13. Massey, V., Palmer, G., Williams, C. H., Swoboda, B. E. P., and Sands, R. H. (1966) Flavin semiquinones and flavoprotein catalysis, in *Flavin and Flavoproteins* (Slater, E. C., ed.), Elsevier, New York, pp. 133–158.
14. Kosower, E. M. (1966) The role of charge-transfer complexes in flavin chemistry and biochemistry, in *Flavin and Flavoproteins* (Slater, E. C. ed.), Elsevier, New York, pp. 1–14.
15. Carrington, A. and McLachlan, A. D. (1967) Introduction to magnetic resonance. Harper and Row Publishing, New York.
16. Hyde, J. S. (1974) Paramagnetic relaxation. *Ann. Rev. Phys. Chem.* **25,** 407–435.
17. Ingram, D. J. E. (1969) Biological and Biochemical Applications of Electron Spin Resonance, Plenum, New York.
18. Hoff, A. J., ed. (1989) Advanced EPR, in *Application in Biology and Biochemistry,* Elsevier, New York.
19. Palmer, G., Müller, F., and Massey, V. (1969) Electron paramagnetic resonance studies of flavoprotein radicals, in *Flavins and Flavoproteins* (Kamin, H., ed.), Elsevier, New York, pp. 123–140.

20. Murataliev, M. B., Klein, M., Fulco, A., and Feyereisen, R. (1997) Functional interactions in cytochrome P450BM3. Flavin semiquinone intermediates, role of NADP(H), and mechanism of electron transfer by the flavoprotein domain. *Biochemistry* **36,** 8401–8412.
21. Hyde, J. S., Eriksson, L. E. G., and Ehrenberg, A. (1970) EPR relaxation of slowly moving flavin radicals: "anomalous" saturation. *Biochim Biophys. Acta.* **222,** 688–692.
22. Hyde, J. S. (1967) Electron nuclear double resonance using an intense nuclear radio frequency field, in *Magentic Resonance in Biological Systems*, Pergamon Press, Oxford, UK, pp. 63–84.
23. Norris, J. R., Thurnauer, M. C., and Bowman, M. K. (1980) Electron spin echo spectroscopy and the study of biological structure and function. *Adv. Biol. Med. Phys.* **17,** 365–416.
24. Iyanagi, T. (1977) Redox properties of microsomal reduced nicotinamide adenine dinucleotide-cytochrome *b5* reductase and cytochrome *b5*. *Biochemistry* **16,** 2725–2730.
25. Nonaka, Y., Fujii, S., and Yamano, T. (1986) The semiquinone state of NADPH-adrenodoxin oxidoreductase in the course of anerobic reduction with NADPH. *J. Biochem.* **99,** 803–814.
26. Peterson, J. A. and Boddupalli, S. S. (1992) P450BM-3: reduction by NADPH and sodium dithionite. *Arch. Biochem. Biophys.* **294,** 654–661.
27. Daff, S. N., Chapman, S. K., Turner, K. L., Holt, R. A., Govindaraj, S., Poulos, T. L., and Munro, A. W. (1997) Redox control of the catalytic cycle of flavocytochrome P-450 BM3. *Biochemistry* **36,** 13,816–13,823.
28. DeRose, V. J., Woo, J. C. G., Hawe, W. P., Hoffman, B. M., Silverman, R. B., and Yelekci, K. (1996) Observation of a flavin semiquinonein the resting state of monoamine oxidase B by electron paramagnetic resonance and electron nuclear double resonance spectroscopy. *Biochemistry* **35,** 11,085–11,091.
29. Lambeth, J. D. and Kamin, H. (1976) Adrenodoxin reductase. Properties of the complexes of reduced enzyme with $NADP^+$ and NADPH. *J. Biol. Chem.* ***251,*** 4299–4306.
30. Narhi, L. O. and Fulco, A. J. (1986) Characterization of a catalytically self-sufficient 119,000-dalton cytochrome P-450 monoxygenase induced by barbiturates in *Bacillus megaterium. J. Biol. Chem.* **261,** 7160–7169.
31. Narhi, L. O. and Fulco, A. J. (1987) Identification and characterization of two functional domains in cytochrome P-450BM-3, a catalytically self-sufficient monooxygenase induced by barbiturates in *Bacillus megaterium. J. Biol. Chem.* **262,** 6683–6690.
32. Masters, B. S. S., Kamin, H., Gibson, Q. H., and Williams, C. H., Jr. (1965) Studies on the mechanism of microsomal triphosphopyridine nucleotide-cytochrome c reductase. *J. Biol. Chem.* **240,** 921–931.

8

Circular Dichroism Studies of Flavoproteins

Andrew W. Munro, Sharon M. Kelly, and Nicholas C. Price

1. Introduction

The aim of this chapter is to illustrate the distinctive contribution that circular dichroism (CD) can make to the study of flavoproteins. In the first section, the underlying principles of CD will be explained, and a brief outline given of the types of information which can be obtained. This is followed by a description of the important practical aspects involved in making reliable CD measurements. The third section will illustrate a number of studies of flavoproteins with examples taken from the recent literature. Finally a number of important general conclusions will be drawn.

1.1. The Underlying Principles of CD

CD refers to the differential absorption of the left and right circularly polarized components of plane polarized radiation. This differential absorption will arise when a chromophore is chiral, either intrinsically by reason of its structure or by being placed in an asymmetric environment (e.g., that generated by the tertiary structure of a protein). The two circularly polarized components are obtained by passage of plane polarized radiation through a *modulator*, typically a quartz crystal subjected to an alternating (50 kHz) electric field. The electric field causes small distortions in the structure of the modulator so that it will transmit each of the two components in turn. The CD instrument (spectropolarimeter) measures the absorption of each circularly polarized component in turn, and then displays the dichroism at a given wavelength as either the difference in absorbance of the two components ($\Delta A = A_L - A_R$), or the ellipticity in degrees (θ) *see* **Notes**. A CD spectrum is obtained when the dichroism is plotted as a function of wavelength. ***(1–3)***. It is important to note that CD can only arise when absorption of radiation occurs; thus it is generally a relatively

From: Methods in Molecular Biology, vol. 131: Flavoprotein Protocols
Edited by: S. K. Chapman and G. A. Reid © Humana Press Inc., Totowa, NJ

straightforward matter to assign a CD band to a particular chemical grouping in a molecule on the basis of the absorption wavelength.

1.2. Information Available from CD

One of the strengths of CD is that a number of aspects of protein structure can be monitored. The far ultraviolet (UV) region (240–190 nm or lower) can be used to give quantitative information on the secondary structure. In this region, the principal absorbing groups are the peptide bonds (the absorption consists of a weak but broad $n \rightarrow \pi^*$ transition centered around 210 nm and an intense $\pi \rightarrow \pi^*$ transition around 190 nm. The usefulness of CD arises from the fact that the different forms of secondary structure show distinct far UV CD spectra (**Fig. 1**). In essence the task of determining the secondary structure of any particular protein consists of analyzing the observed spectrum in terms of the spectra of a reference set of proteins for which structures have been determined to high resolution by X-ray crystallography.

Analysis of far UV CD spectra can be complicated by a number of factors including the contribution that other chromophores (including aromatic amino acids and disulphide bonds) can make in this region *(4–6)*.

The near UV CD spectrum of a protein (320–250 nm) arises predominantly from the aromatic amino acids (phenylalanine, tyrosine, and tryptophan), although the disulphide bond can also contribute. Small model compounds of the aromatic amino acids exhibit CD spectra because the aromatic chromophore is linked to the nearby chiral α-carbon atom. The tertiary structure of a protein will place these amino acids in a variety of asymmetric environments. The intensities of aromatic CD bands are influenced by the rigidity of the protein with the more highly mobile side chains having lower intensities and by interactions between aromatic side chains, which can be significant if they are close (within 1 nm). The contributions of individual side chains to the overall spectrum can be investigated by examining the spectra of suitable mutant derivatives in which the amino acids are replaced in turn *(7)*.

The CD spectrum in the region where the nonprotein chromophore absorbs can be used to give information on the environments of these cofactors. In the context of flavoproteins, CD signals from the flavin chromophores (generally flavin mononucleotide (FMN) or flavin adenine dinucleotide (FAD)) may be observed over the normal visible absorbance range for this cofactor, i.e., between 300–500 nm in the oxidized form. The flavin isoalloxazine ring system can be reduced by a single electron (to the semiquinone form) or two electrons (to the hydroquinone). The hydroquinone form has little absorbance in the visible region and hence minimal visible CD; but the semiquinone forms (both neutral and anionic) usually have distinctive visible *absorbance spectra* and may also be amenable to study by CD. Whereas the *absorbance spectra* of

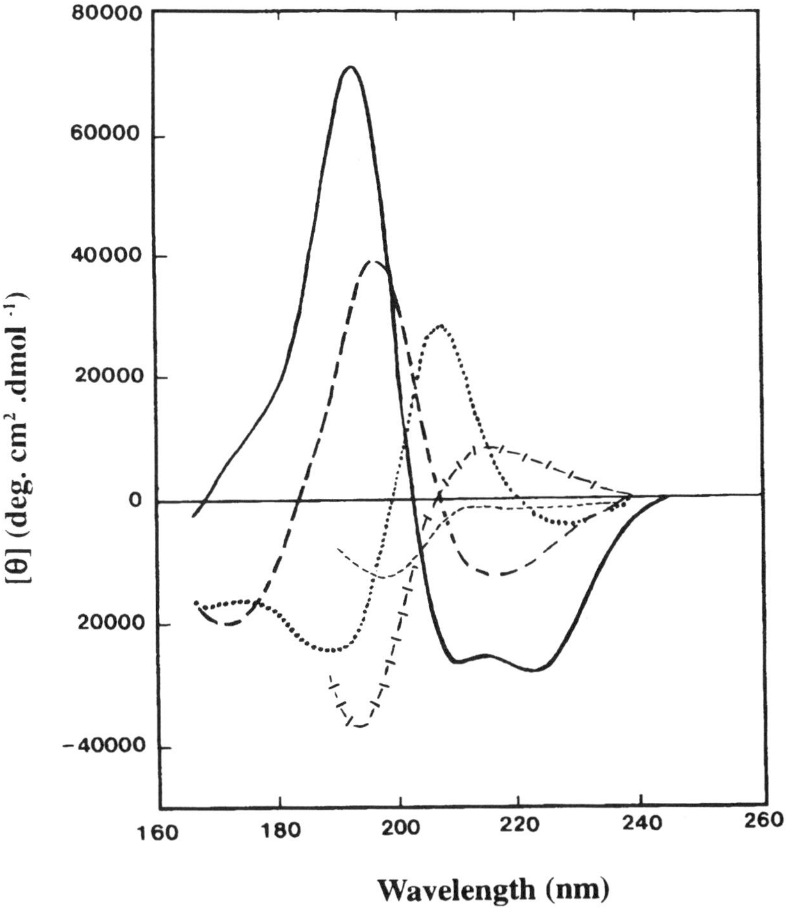

Fig. 1. Far UV CD spectra associated with various types of secondary structure: solid curve, α-helix; long dashes, antiparallel β-sheet; dots, type I β-turn; dash-bar-dash, left-handed extended 3_{10} helix; short dashes, irregular structure.

anionic semiquinones usually show strong overlaps with the spectra of the oxidized flavin, neutral semiquinones generally absorb at longer wavelengths (550–650 nm), allowing their CD properties to be studied without interference from the other forms.

In summary, the far UV CD can be used to give quantitative information on secondary structure. The near UV and visible CD tend to be used as indices of the tertiary structure and the binding site for the cofactor, respectively. CD in the near UV and visible regions is of particular value when a family of related proteins is being examined, e.g., a range of site directed mutants, since relatively small differences in the structures of the proteins can be detected. This can be of major benefit in the study of flavoenzymes, allowing the analysis of the flavin and protein environments.

Changes in the CD (in any of the spectral regions) can be used to monitor structural changes in the protein which might accompany ligand binding, or could be brought about by the action of heat or a denaturing agent. The rate of such structural changes can be deduced by measuring the time dependence of CD signals. This approach has been used extensively to measure the rate of refolding of an unfolded protein, after the process has been initiated by dilution of the denaturant *(8,9)*. By measuring CD signals in different spectral regions, it is possible, for instance, to assess whether a cofactor binding site in a protein is more sensitive to the denaturant than the overall secondary structure. Thus, the decrease in the intensity of the ellipticity in the far UV region can be used to measure the disruption of secondary structure, while that in the visible region can be used to assess the integrity of the binding site of the flavin cofactor. Free flavins exhibit relatively small visible CD and thus the dissociation of flavin from the protein should lead to an almost complete loss of this signal. However, at intermediate temperature or denaturant concentration, alterations in the flavin environment may be detected by changes in the visible CD signal.

2. Experimental Aspects

In this section some of the more important experimental aspects of CD will be outlined so that the reader can appreciate the conditions under which reliable data can be obtained.

2.1. Instrumentation

The light source for most CD instruments is a xenon arc, which gives a good output over the range of wavelengths used for virtually all CD studies on proteins. The spectropolarimeter is flushed with nitrogen gas in order to remove oxygen from the lamp housing and the sample compartment. This not only prevents ozone formation, but allows measurements to be made below 200 nm (where oxygen absorbs). The spectropolarimeter should be regularly maintained and calibrated with a suitable standard such as 1*S*-(+)-10-camphorsulphonic acid *(1,10)*.

In most work with biological samples, the observed signals in CD are very small, thus the observed difference in absorbance of the two circularly polarized components is of the order of 3×10^{-4}, corresponding to an ellipticity of 10 millidegrees. This means that the results of a large number of individual measurements (which are made at the frequency of the alternating electric field, i.e., 50 kHz) are accumulated in order to increase the signal to noise ratio. A number of experimental parameters can be adjusted to improve the reliability of the data. These include the time constant, the scan rate, the number of scans and the bandwidth. For "steady-state" CD measurements, typical conditions would be a time constant of 2 s, a scan rate of 10 nm/min, accumulation of 2–4 scans and a bandwidth of 2 nm or less ***(1,2)***. When kinetic measurements are undertaken on the stopped-flow timescale (tens of milliseconds), a very short time constant is used, but it is usually necessary to accumulate the results of a large number of runs to produce an acceptably low level of noise.

2.2. Sample

It is important that the sample should be homogeneous and freed of highly scattering particles either by centrifugation or passage through a suitable filter (0.2 μm). The total absorbance of the sample should not exceed about one unit, otherwise the spectral noise will become excessive as an increasingly high voltage is applied to the photomultiplier. Above a certain voltage, an automatic cut-off may come into operation leading to an apparent decline of the CD signal to zero. It is essential to minimize the absorption due to other components in the mixture (buffers, electrolytes, detergents etc.), so that the absorbance limit is not exceeded. Most problems arise in the far UV region; suitable buffer systems that can be used down to 190 nm (or even lower) include phosphate, borate and low molarity (≤ 20 m*M*) Tris. High concentrations of chloride ions (> 50 m*M*) should be avoided if possible; fluoride and sulphate are more suitable anions to maintain ionic strength. In the near UV and visible regions, buffers are unlikely to cause problems of this type, but it is always good practice to check the characteristics of a blank solution (i.e., with no protein) in each case, before undertaking detailed studies.

2.3. Data Presentation and Analysis

For far UV CD, where the peptide bond is the principal absorbing species, it is the molar concentration of these bonds that is required. This is obtained by dividing the concentration in mass terms (i.e., mg/mL) by the mean residue weight (MRW). The MRW is obtained by dividing the molecular mass by $N - 1$, where N is the number of amino acids in the polypeptide chain. For most proteins, the MRW is usually close to 110.

When CD data are expressed in terms of absorbance, the units are those of the difference in molar absorbance $\Delta\varepsilon = \varepsilon_L - \varepsilon_R$, i.e., M^{-1} cm^{-1}. When expressed in terms of ellipticity, the mean residue ellipticity ($[\theta]_{mrw,\lambda}$ at a given wavelength, λ) is quoted in units of deg·cm^2·dmol^{-1}, and is given by the equation:

$$[\theta]_{mrw,\lambda} = MRW \cdot \theta / 10 \cdot d \cdot c$$

where θ is the observed ellipticity (in degrees) at the wavelength of interest, d is the pathlength (cm) and c is the concentration (expressed in units of g/mL). The numerical relationship between the two sets of units is:

$$[\theta]_{mrw} = 3298\ \Delta\varepsilon$$

It is important to measure the concentration of protein accurately (preferably within ± 5%) if reliable estimates of secondary structure are to be made from far UV measurements. Amino acid analysis (for pure protein samples) provides an excellent quantitative method, but the common dye-binding and chromogenic techniques (e.g., ***11–13***) may also be appropriate, providing that the protein used as the standard is sufficiently similar in behavior to the protein of interest.

Some workers have expressed reservations as to whether mean residue ellipticity is an appropriate unit for expressing the amplitude of the near UV and visible CD signals of a protein, since only a small number of aromatic amino acids or cofactors contribute to the CD signals in these regions. An alternative is to express the data in terms of the molar concentration of the intact protein rather than of the repeating peptide unit. It is clearly important in any study that the molar basis of the units is clearly stated.

2.4. Amount of Protein Required

The amount of protein required for CD measurements can be judged from the need to keep the absorbance less than about one unit. Typical cell pathlengths for far UV CD work are in the range 0.01–0.05 cm and protein concentrations are in the range from 0.2–1 mg/mL. Depending on the design of the cell, the volume of sample required can range from about 1 mL to as little as 50 µL (recovery of sample can be very difficult with the smaller volume cells). Thus, an acceptable far UV CD spectrum can be obtained with as little as 10 µg protein, but in general 100–500 µg is required to explore the optimum conditions for recording spectra. A range of concentrations may be studied to obtain a smooth spectrum of maximal intensity within the far UV range (usually from 190–260 nm). Data down to below 200 nm is required for accurate secondary structure analysis and the protein sample should be adjusted to a suitable concentration to allow collection of data to this wavelength. The CD signals in the near UV and visible regions of flavoproteins are considerably weaker than those in the far UV, as a consequence of the much lower molar concentrations of the relevant chromophores (i.e., aromatic amino acids and flavins, respectively) compared with that of the peptide bonds. For measurements in these spectral regions, typical protein concentrations would be in the range 0.5–2 mg/mL and a cell pathlength of 0.5–2 cm; the volume required would be typically in the range 1–5 mL. Thus, the amounts of protein required for these experiments will be of the order of several mg. For the oxidized flavin chromophore in the visible region (typically 320–600 nm), the quantity of protein required reflects the flavin content (one or more flavins) and the extent of the chirality. The molar CD signal of a particular protein-bound flavin can

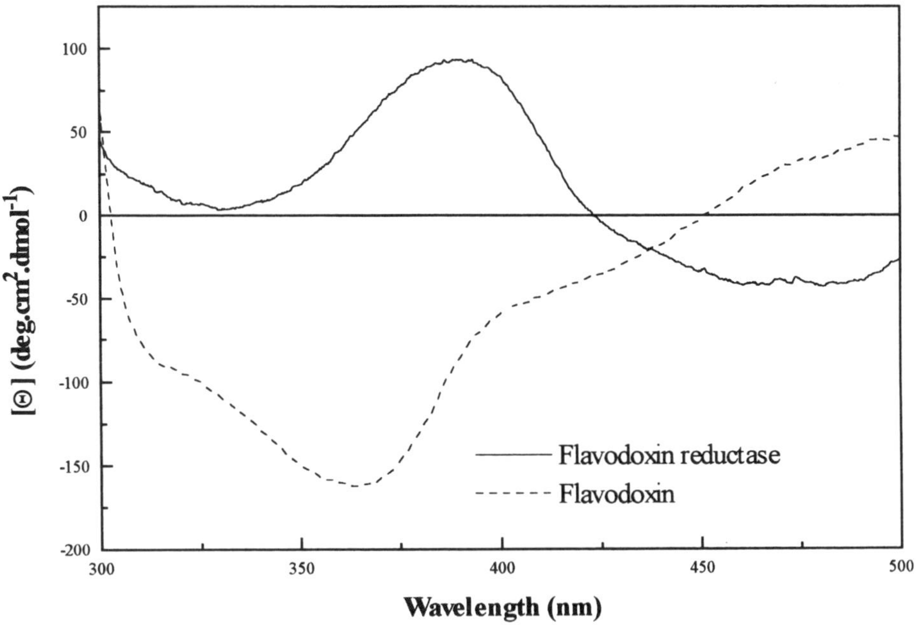

Fig. 2. Visible region CD of *E. coli* flavodoxin (dashed line) and flavodoxin reductase (solid line). Both spectra were recorded in 0.5 cm pathlength cells at protein concentrations of 1.5 mg/mL, corresponding to 54.3 μ*M* flavodoxin reductase and 76.5 μ*M* flavodoxin.

be difficult to predict. Indeed it should be remembered that a CD signal may be either positive or negative, so that in a multiflavin protein or system, there could be a certain amount of "canceling out" of signals. For example, the FMN-containing *E. coli* flavodoxin shows negative ellipticity over the range of approx 310–450 nm, with a maximal visible CD signal at 365 nm (–160 deg $\cdot$ cm^2 $\cdot$ dmol^{-1}). By contrast, its redox partner flavodoxin reductase exhibits positive ellipticity over the range 300–420 nm, with a maximum at approx 390 nm (+95 deg $\cdot$ cm^2 $\cdot$ dmol^{-1}) (**Fig. 2**).

An important factor to be borne in mind during the analysis of the visible CD of flavoproteins is the strength of binding of the flavin. For instance, the spectrum of a 50 kDa flavoprotein might be recorded at a concentration of 0.5 mg/mL (i.e., 10 μ*M*). If the flavin is not covalently linked to the protein and is rather weakly bound (e.g., the FMN moiety of flavocytochrome P-450 BM3, *see* **Subheading 3.**) then a significant proportion of the flavin may not be bound to the protein and thus may not contribute to the visible CD spectrum. However, for proteins such as bacterial flavodoxins the K_d for the FMN is in the nanomolar range and dissociation will be minimal at the above concentration ***(14)***.

3. Examples of Studies on Flavoproteins Using CD

A number of flavoproteins have been analyzed by CD spectroscopy. Detailed studies of the flavin environment and its perturbation during protein unfolding have been undertaken with flavocytochrome P-450 BM3 from *Bacillus megaterium*, a fatty acid hydroxylase enzyme containing 1 molecule each of FAD, FMN, and haem ***(15)***. This enzyme and each of its component domains (heme, diflavin, FAD, and FMN) have been overexpressed in *E. coli* and analyzed by CD. Far UV CD studies showed that the secondary structures of the individual haem (P-450) and diflavin (P-450 reductase) domains are the same as in the intact flavocytochrome. However, the CD signals due to the flavin and haem groups were considerably different between the individual domains and the intact protein, reflecting altered environments of these cofactors. Because the activity of the isolated domains when added together is >100-fold less than that of the intact enzyme, it is reasonable to propose that in the mixture there has been some disruption to the essential interactions between cofactors that contribute to the high activity of the intact flavocytochrome ***(16)***. More recently, similar alterations in far and near UV CD and visible CD have been observed on the expression of the individual FMN-(flavodoxin-like) and FAD-(NADP$^+$-flavodoxin reductase-like) containing domains of P-450 BM3 (unpublished data). Also with P-450 BM3, CD was used to compare the structural stability (far UV and near UV CD) with the stability of binding of flavins (visible CD). The disappearance of one of the two visible CD bands was seen to occur over the range of denaturant (guanidinium chloride, GdnHCl) from 0.5 – 1 *M* at which point all of the more weakly bound FMN was lost from the enzyme. Between 1 *M* and 2 *M* GdnHCl, all FAD was also lost, along with the second visible CD band (**Fig. 3**). All FMN was lost from the protein at a point where only very minor changes in secondary structure were observed; the results suggested that flavin loss represented a distinct step in the unfolding process of the enzyme ***(17)***.

In studies of wild-type and mutant forms of the FAD-containing lipoamide dehydrogenase from *Azotobacter vinelandii*, it was shown that *N*-terminal deletion mutants (Δ9 and Δ14 amino acids) had lower β-sheet structural content compared with wild type. Although the Δ14 mutant had similar α-helical content to the wild type when measured at 30 μ*M*, dilution of the enzyme led to a dramatic decrease in the α-helix content, which was ascribed to a weakening of intersubunit interactions in this mutant. Moreover, distinct changes in the visible CD of the mutants were observed. These included the development of a negative CD signal in the region of the longer wavelength absorption band of the FAD in the Δ14 mutant, reflecting considerable alteration of the flavin microenvironment ***(18)***.

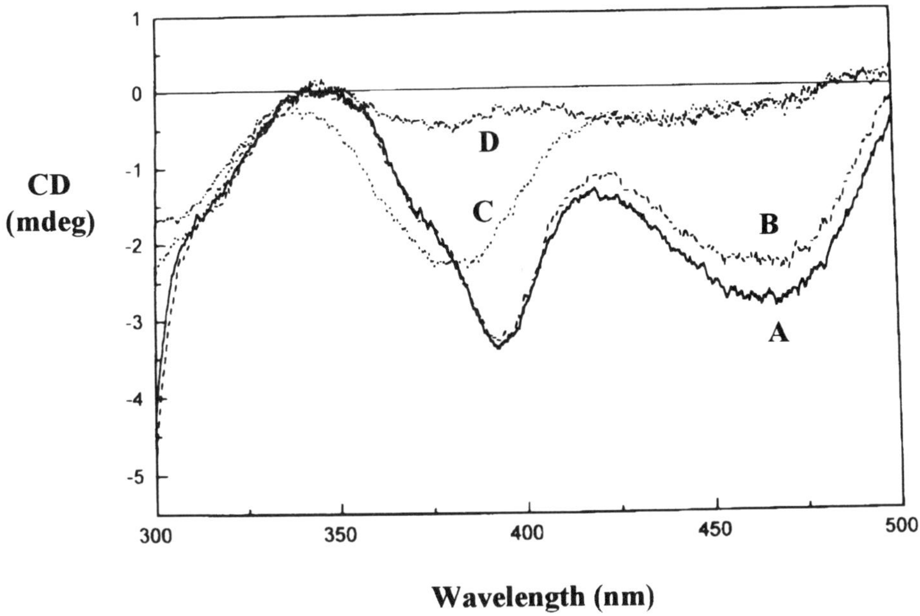

Fig. 3. Visible region CD spectra of the reductase (FMN- and FAD-containing) domain of flavocytochrome P-450 BM3 (40 µ*M*) at various concentrations of guanidinium chloride (GdnHCl). Spectra were taken after the enzyme was incubated in 20 m*M* MOPS (7.4) + 100 m*M* KCl containing GdnHCl at 0 *M* (A), 0.5 *M* (B), 1 *M* (C) and 3 *M* (D). All the FMN is dissociated from the enzyme by 1 *M* (C) and one of the two visible CD bands disappears. All the FAD is removed by 3 *M* (D) and the visible CD signal is almost completely bleached ***(17)***.

Similarly, CD studies of pathogenic variants T266M and G116R of the human FAD-binding electron transferring flavoprotein (ETF), which are associated with the inherited metabolic disease glutaric acidaemia (type II), revealed differences between the structures of the wild-type and mutant forms. The mutant G116R folded in an aberrant fashion to give an inactive, unstable form. In the case of the mutant T266M, the far UV CD indicated that folding was similar to that of the wild-type protein; however the visible CD indicated marked differences in the flavin environment. These were ascribed to the loss of a hydrogen bond at N5 of the flavin ring, which also resulted in a 10-fold stabilization of the FAD semiquinone form ***(19)***.

Flavins are often found in enzymes along with one or more haem groups (as in P-450 BM3 or the tetra-haem flavocytochrome c_3 fumarate reductase from *Shewanella* NCIMB 400) or other cofactors. Because the haem groups absorb in the same spectral region as flavins but with extinction coefficients generally around 10-fold higher than those of flavins, it can be difficult to perform spectrophotometric analysis of flavins in such proteins. There may be particular merit in the use of visible CD as a spectroscopic probe of flavin in such enzymes. A good example is the nitrate reductase from *Chlorella,* where the flavin contribution is masked in the visible absorption spectrum of this molybdopterin- and haem *b*-containing enzyme. By contrast, strong visible CD signals from FAD can be observed. Positive CD signals from the FAD are located at 311 and 387 nm, respectively, with negative CD signals at 460 nm and 487 nm. These features were lost on reduction ***(20)***. Strong features are also seen for the FMN in the flavocytochrome b_2 lactate dehydrogenases from *Hansenula anomala* and *Saccharomyces cerevisiae*, with a maximum at ~ 325 nm, a broad minimum at ~ 375 nm, and a particularly intense negative band in the near UV CD at 270 nm. This band disappears on flavin removal and is likely to reflect stacking interactions between the FMN and two tyrosine side chains ***(21)***.

Circular dichroism has also been used to monitor the binding of flavins to apoproteins and of flavoprotein holoenzymes with other proteins. In the case of lipoamide dehydrogenase (a component of α-keto acid dehydrogenase complexes) from pig heart, the binding of FAD was found to induce a change in the weak, negative CD signal at 370 nm (due to free FAD) to a strong, positive one at the same wavelength ***(22)***. In an elegant series of experiments, D'Anna and Tollin demonstrated that the binding of FMN to apoprotein flavodoxins from *Peptostreptococcus elsdenii, Desulfovibrio vulgaris,* and a variety of other bacteria induced large changes in the far UV CD indicative of increased protein secondary structure ***(23)***. By contrast, an increase in far UV CD intensity (indicating greater secondary structural content) was observed for FAD-containing lipoamide dehydrogenase during dissociation of FAD ***(24)***. With yeast flavocytochrome b_2, no net effect on protein structure was observed when FMN was removed ***(25)***. D'Anna and Tollin also investigated effects in the visible CD and showed that FMN-binding resulted in a change of sign and an approximately 10-fold increase in the intensity of the flavin signal compared with free FMN. On the basis of the appearance of the visible CD spectra of different microbial flavodoxins, it was possible to define two broad classes based on whether the crossover point from negative-to-positive ellipticity occurred at ~400 nm (*Clostridium pasteurianum* type) or at ~440 nm (*Rhodospirillum rubrum* type). The distinction seems to reflect a specific difference between the nature of interaction of the FMN with the protein in the two classes. Sakamoto and co-workers ***(26)*** have used CD as a probe for protein-protein interaction by monitoring changes in the visible CD spectrum of the FAD-containing adrenodoxin reductase (ADR) when it was titrated against a solution of its redox partner, the iron-sulphur protein adrenodoxin (AD). The maximal change in the visible flavin CD signal was observed at 452 nm when the sum of the spectra of the individual AD and ADR components were subtracted from that of the AD/ADR mixture. From the plot of the change in CD at 452 nm against the concentration of ADR added, it was possible to calculate a K_d of 4.56×10^{-7} *M* ***(26)***.

4. Guidelines for Successful CD Analysis of Flavoproteins

CD spectroscopy permits the study of a number of aspects of the structure of flavoproteins. The secondary structure of the flavoprotein can be examined by far UV CD, and the spectrum of wild-type protein provides a reference point for the comparison of the gross conformation of mutant enzyme forms. Comparisons in the near UV and visible regions may be more informative, highlighting alterations in the flavin environment which are not necessarily associated with large overall structural perturbations ***(18)***. For flavocytochromes and flavoproteins with additional prosthetic groups, visible CD can be a useful way of examining the properties of the flavin(s) in isolation. The visible absorption spectra of haem groups, in particular, mask those of flavins, but the visible CD of protein-bound flavins may be as large (or larger) than those of haems. Perturbations in the far UV, near UV and visible CD of flavoproteins can also be used to examine their interactions with protein partners.

For reliable CD results, the most important practical points to note are:

1. The CD instrument should be calibrated regularly.
2. The experimental parameters (time constant, scan rate, number of scans) should be appropriate to the measurement.
3. The flavoprotein preparation should be pure and its concentration accurately known. The concentration used should be appropriate in terms of absorbance in the spectral region of interest. Buffers used should not absorb strongly in this region.
4. The state of association of flavin with the protein should be checked under the conditions employed.

5. Notes

The term ellipticity arises from the fact that if the two components were recombined after passage through the sample, the vector of the resulting radiation would trace out an ellipse. The ellipticity, θ, is defined as $\tan^{-1}(b/a)$ where *b* and *a* are the minor and major axes of the ellipse, respectively. There is a simple numerical relationship between ΔA and θ (in degrees), that is, $\theta = 32.98\Delta A$.

References

1. Martin, S. R. (1996) Circular dichroism, in *Proteins Labfax* (Price, N. C., ed.), Bios Scientific Publishers Oxford, UK, and Academic Press, San Diego, CA, pp. 195–204.

2. Kelly, S. M. and Price, N. C. (1997) Application of circular dichroism to studies of protein folding and unfolding. *Biochim. Biophys. Acta* **1338,** 161–185.
3. Fasman, G. (ed.) (1996) *Circular Dichroism and the Conformational Analysis of Biomolecules,* Plenum Press, New York, 788 pp.
4. Woody, R. W. (1995) Circular dichroism. *Methods Enzymol.* **246,** 34–71.
5. Kosen, P. A., Creighton, T. E., and Blout, E. R. (1981) Circular dichroism spectroscopy of bovine pancreatic trypsin inhibitor and five altered conformational states. Relationship of conformation and the refolding pathway of the trypsin inhibitor. *Biochemistry* **20,** 5744–5760.
6. Chaffotte, A. F., Guillou, Y., and Goldberg, M. E. (1992) Kinetic resolution of peptide bond and side chain far-UV circular dichroism during the folding of hen egg white lysozyme. *Biochemistry* **31,** 9694–9702.
7. Freskgård, P.-O., Mårtensson, L.-G., Jonasson, P., Jonsson, B.-H., and Carlsson, U. (1994) Assignment of the contribution of tryptophan residues to the circular dichroism spectrum of human carbonic anhydrase II. *Biochemistry* **33,** 14,281–14,288.
8. Kuwajima, K., Yamaya, Y., Miwa, S., Sugai, S., and Nagamura, T. (1987) Rapid formation of secondary structure framework in protein folding studied by stopped-flow circular dichroism. *FEBS Lett.* **221,** 115–118.
9. Elöve, G. A., Chaffotte, A. F., Roder, H., and Goldberg, M. E. (1992) Early steps in cytochrome *c* folding probed by time-resolved circular dichroism and fluorescence spectroscopy. *Biochemistry* **31,** 6876–6883.
10. Johnson, W. C. (1990) Protein secondary structure and circular dichroism—a practical guide. *Proteins: Struct. Funct. Genet.* **7,** 205–214.
11. Bradford, M. M. (1976) A rapid and sensitive method for the quantitation of microgram quantities of protein utilizing the principle of protein-dye binding. *Anal. Biochem.* **72,** 248–254.
12. Smith, P. K., Krohn, R. I., Hermanson, G. T., Mallia, A. K., Gartner, F. H., Provenzano, M. D., Fujimoto, E. K., Goeke, N. M., Olson, B. J., and Klenk, D. C. (1985) Measurement of protein using bicinchoninic acid. *Anal. Biochem.* **150,** 76–85.
13. Lowry, O. H., Rosebrough, N. J., Farr, A. L., and Randall, R. J. (1951) Protein measurement with the Folin phenol reagent. *J. Biol. Chem.* **193,** 265–275.
14. Mayhew, S. G. and Tollin, G. (1992) General properties of flavodoxins, in *Chemistry and Biochemistry of Flavoenzymes,* vol. III (Müller, F., ed.), CRC Press, Boca Raton, FL; Ann Arbor, MI; London, pp. 389–426.
15. Narhi, L. O. and Fulco, A. J. (1986) Characterization of a catalytically self-sufficient 119,000-Dalton cytochrome P-450 monooxygenase induced by barbiturates in *Bacillus megaterium. J. Biol. Chem.* **261,** 7160–7169.
16. Munro, A. W., Lindsay, J. G., Coggins, J. R., Kelly, S. M., and Price, N. C. (1994) Structural and enzymological analysis of the interaction of isolated domains of cytochrome P-450 BM3. *FEBS Lett.* **343,** 70–74.
17. Munro, A. W., Lindsay, J. G., Coggins, J. R., Kelly, S. M., and Price, N. C. (1996) Analysis of the structural stability of the multidomain enzyme flavocytochrome P-450 BM3. *Biochim. Biophys. Acta* **1296,** 127–137.

18. Visser, A. J. W. G., Van Berkel, W. J. H., and De Kok, A. (1995) Changes in secondary structure and flavin microenvironment between *Azotobacter vinelandii* lipoamide dehydrogenase and several deletion mutants from circular dichroism. *Biochim. Biophys. Acta* **1229,** 381–385.
19. Salazar, D., Zhang, L. N., de Gala, G. D., and Frerman, F. E. (1997) Expresssion and characterization of two pathogenic mutations in human electron transfer flavoprotein. *J. Biol. Chem.* **272,** 26,425–26,433.
20. Vanoni, M. A., Curti, B., and Zanetti, G. (1992) Glutamate synthase, in *Chemistry and Biochemistry of Flavoenzymes,* vol. III (Müller, F., ed.), CRC Press, Boca Raton, FL, pp. 309–317.
21. Sturtevant, J. M. and Tsong, T. Y. (1969) Yeast L-lactate dehydrogenase (cytochrome b_2). VI. Circular dichroism of the holoenzyme. *J. Biol. Chem.* **244,** 4942–4950.
22. Brady, A. H. and Beychok, S. (1969) Optical activity and conformation studies of pig heart lipoamide dehydrogenase. *J. Biol. Chem.* **244,** 4634–4637.
23. D'Anna Jr., J. A. and Tollin, G. (1972) Studies of flavine-protein interaction in flavoproteins using protein fluorescence and circular dichroism. *Biochemistry* **11,** 1073–1080.
24. Veeger, C., Voetberg, H., Visser, J., Staal, G., and Koster, J. (1971) Conformational transitions in flavoproteins, in *Flavins and Flavoproteins* (Kamin, H., ed.), University Park Press, Baltimore, MD, pp. 261–264.
25. Tsong, T. Y. and Sturtevant, J. M. (1969) Yeast L-lactate dehydrogenase (cytochrome b_2). V. Circular dichroism of the flavin mononucleotide-free apoenzyme. *J. Biol. Chem.* **244,** 2397–2402.
26. Sakamoto, H., Ichikawa, Y., Yamano, T., and Takagi, T. (1981) Circular dichroic studies on the interaction between reduced nicotinamide adenine dinucleotide phosphate-adrenodoxin reductase and adrenodoxin. *J. Biochem.* **90,** 1445–1452.

9

Vibrational Spectroscopy of Flavoproteins

Iain D. G. Macdonald

1. Introduction

Several techniques under the broad umbrella of "vibrational spectroscopy" have aided in the identification of structural and electronic factors associated with the catalytic mechanisms of many biological systems *(1)*. A complete elucidation of the mechanism of function of a biological center requires knowledge of the exact nature of bond strengths and their dependence upon the protein environment. By probing the intrinsic vibrational properties of species of interest, which are almost invariably prosthetic group(s), one can obtain information on interactions as diverse as molecular recognition events *(2)* and mechanistic intermediates *(3)* to dynamics associated with ligand/substrate binding *(4)* and protein motions *(5)*. Flavin-containing proteins are a vital component of numerous enzyme systems, participating in a myriad of functions within organisms. Their roles range from drug detoxification to aromatic acid synthesis *(6,7)*. Several spectroscopic techniques have been applied to studying the molecular properties of flavoproteins in order to understand more fully their functional basis as efficient electron donors/acceptors. In this chapter one shall deal exclusively with what vibrational spectroscopy has provided and can offer towards our understanding of the inter-relationships between molecular properties and flavoprotein function. A brief outline of theory and instrumentation is followed by an extensive experimental protocol for the associated techniques. This includes sample preparation for the flavoprotein, generation of small metal particles required for surface enhanced resonance Raman scattering (SERRS) studies and the formation of semiquinoid species. To conclude, data interpretation and its relevance to flavoprotein function is discussed.

From: Methods in Molecular Biology, vol. 131: Flavoprotein Protocols
Edited by: S. K. Chapman and G. A. Reid © Humana Press Inc., Totowa, NJ

1.1. Theory and Instrumentation

The vibrational spectroscopic techniques of Raman scattering and infrared spectroscopy have been used extensively for probing biological systems. However, strong background absorption by the media and a lack of specificity has limited the applicability of infrared spectroscopy to only a few favorable cases e.g., CO-bound hemes *(4)*. On the other hand, Raman scattering becomes a highly sensitive and selective *in situ* technique when the incident excitation frequency becomes coincident with an electronic transition of the molecular species of interest. This is known as the resonance Raman effect (RR) and can result in an increase in scattering efficiency of 10^6 *(8)*. Additionally, adsorption of proteins onto small metallic particles provides further scattering enhancement of the order of 10^3–10^4 above that gained from RR, thereby enabling very low concentrations to be probed. In addition, this technique, known as surface enhanced resonance Raman scattering [SERRS], can provide surface orientation information on self-assembled protein monolayers *(9)*. Consequently, Raman techniques are the chosen method by researchers for the study of flavoproteins.

How do these researchers go about acquiring data? Briefly, an excitation laser source is focused onto the sample. Scattered light is collected at 90° or 180° to the incident beam and the light (which has its polarization scrambled) is dispersed via a monochromator onto the detector. Rayleigh scattering is removed by an interference filter which is placed before the scattered light collection optics. Computer interfacing and processing combined with appropriate software allow for remote control and manipulation of data collection. Typically, modern detectors are charged coupled devices which offer improved sensitivity and enable sampling of all the light in real time. Experiments are generally carried out between 275 and 298 K with low laser power (5–50 mW) and protein concentrations from submicromolar (SERRS studies) to millimolar quantities (RR). Calibration of Raman frequency typically uses well characterized compounds such as indene.

2. Materials

2.1. RR Scattering

1. It is highly desirable to use fresh, pure flavoprotein, which has been stored at –70°C typically in a 20% (v/v) glycerol buffer.
2. Sodium/potassium phosphate buffers (Aldrich, Milwaukee, WI; Fisons, Loughborough, UK; Sigma Chemical Co., St. Louis, MO: > 99.0% purity] should be prepared to the correct pH and ionic strength in distilled water. Buffer conditions are typically 50–100 m*M* phosphate in the pH range of 7.0–7.5. It is important to note that if TRIS (Tris[hydroxmethyl]aminomethane) (Aldrich/Fisons/Sigma:

> 99.0% purity) buffer is used it should be prepared *fresh* to eliminate the possibility of cell/organism growth within the media.
3. A clean capillary or nuclear magnetic resonance (NMR) tube or fluorescence cell is required for the sample.

2.2. SERRS

2.2.1. Colloidal Metal

1. Concentrated hydrochloric acid (Fisher Scientific, Pittsburgh, PA, 36.7%, ACS Plus Grade) and concentrated nitric acid (Fisher Scientific 69.8% ; ACS Plus Grade) are required to create a highly oxidizing mixture to clean experimental glassware. *Aqua regia*, as this mixture is known, consists of one part concentrated nitric acid added to three parts concentrated hydrochloric acid. Note that the preparation and storage of *aqua regia* should take place in a well vented fumehood and handlers should wear heavy duty gloves, laboratory coat and safety glasses.
2. Glassware required includes a wide necked 1 L round-bottomed flask and a glass stirring rod with glass links. A stainless steel white nylon bristled brush, which is chemically resistant, is used to clean only colloidal glassware.
3. Silver nitrate (light sensitive, Johnson Matthey 99.9999% purity),
4. Trisodium citrate (Aldrich/Fisons/Sigma: > 99.0% purity)
5. Sodium borohydride (Aldrich/Fisons/Sigma: > 99.0% purity)
6. Solutions 3.–5. are prepared freshly by dissolving in distilled water. The "good" colloid should be stored in the dark and is stable for 2–3 mo.

2.2.2. Sample Preparation

1. Fresh, pure flavoprotein is required.
2. Sodium/potassium phosphate buffer salts (Aldrich/Fisons/Sigma: > 99.0% purity) are prepared in distilled water.
3. Typically a 100 m*M* phosphate solution in the range pH 5.8–7.2 has been used in these experiments.
4. Fresh L-ascorbic acid (Sigma > 99.0%), a 1% w/v solution is prepared in distilled water.
5. If required, a poly-L-lysine [4 μ*M*] or 0.8 m*M* spermine (Sigma > 99.0% purity, store at –20°C) solution is freshly prepared in distilled water.
6. Sample cells usually are a capped fluorescence cell or capillary tube or microtiter plate.

2.3. Semiquinone Formation

1. Fresh, pure flavoprotein is required.
2. Sodium/potassium phosphate or TRIS buffers (Aldrich/Fisons/Sigma: > 99.0% purity) with an ionic strength of 100 m*M* and pH range of 7.0–7.6 have been used.
3. EDTA (Sigma/Fisons/Aldrich 99.0%; 0.7–50 m*M*) and exogenous FMN (Sigma > 99.0%; 1.5–40 μ*M*), dissolved in buffer, are used for chemical reduction in conjunction with a light source e.g., UV filtered lamp or laser.

4. If natural reductants are used i.e., freshly prepared pyruvate with sodium L-lactate (10–50 m*M*) or 0.5–3 m*M* NADPH (all compounds from Sigma 98.0% purity, store at –20°C). These would be placed in the sample cell after preparation in the desired buffer. Sample cells used include a Thunberg cell or capillary tube.

3. Methods

It is important to realize that the purity and integrity of the flavoprotein has to be established before commencing spectroscopic measurements using any of the above techniques. Many interferents can contribute to the Raman scattering process and therefore, in order to allow for clear and valid interpretation and relate to enzymatic function, the status of the sample before and, if possible, after study should be known. Flavoproteins should be prepared and activities measured according to the protocols for each individual case.

3.1. RR Scattering

1. An aliquot of flavoprotein should be thawed slowly by placing on ice.
2. An aliquot of the thawed flavoprotein is checked for purity and activity, most commonly by UV/visible absorption measurements. Determination of activity assays and purity levels of flavoproteins have been discussed in Chapters 1, 2, 5, and 11, e.g.
3. Buffer and protein should be mixed to the desired protein concentration and volume. Typical concentrations for oxidized flavin RR studies are in the range 100 μ*M*–1.5 m*M* with sample volumes of ~10–500 μL. The sample should be allowed to equilibrate at the desired temperature. This may take up to 20–30 min.
4. Sample is carefully transferred by a precleaned pipet or Hamilton syringe (Hamilton Co., Reno, NV) to the capillary tube or fluorescence cell, mounted to a fixed geometrical arrangement and kept at the desired experimental temperature. Temperatures can be maintained either by circulating coolant or directing cooled gas towards the sample.
5. The sample is now ready to be spectroscopically probed. Typically, argon ion (363.8 nm, 457.9 nm, 488 nm, 514.5 nm), krypton ion (413.1 nm, 647.1 nm), He/Cd (441.6 nm) and He-Ne (632.8 nm) lasers are used to provide a variety of excitation energies. Selectivity for a species of interest i.e., a charge transfer complex between flavin and substrate, or minimization of fluorescence (which is a major interferent in resonance Raman studies) are important factors in deciding the excitation line. Remember that selectivity arises on fulfilling a resonance condition and species are probed with energies at or near their electronic absorption characteristics. "Tweaking" of the micropositioners for both focusing and collection optics maximizes the efficiency of the detected scattering signal, Typically, background fluorescence and cosmic "spikes" are removed by software packages such as supplied by Renishaw Instruments or Galactic Industries (Salem, NH).
6. If possible, after the experiment is completed the sample should be checked (generally by optical methods) for sample degradation.

3.2. SERRS

The highly sensitive and selective technique of SERRS requires adsorption of the flavoprotein onto surfaces that possess "atomic roughness" characteristics. Discrete metallic particles/arrays, which are smaller than the excitation wavelength used (i.e., nanometer scale) are required for an enhancement in scattering efficiency ***(10)***. To be successfully applied, it is critical that the correct "morphology" and size of particle be generated. The most common form of particles used are metal colloids although it should be noted that atomic roughness can be created on electrodes using multiple oxidation/reduction cycles ***(11)***. Silver has been the metal of choice for several researchers due to its intrinsic dielectric properties which allow it to provide the greatest enhancement of Raman scattering. However, gold surfaces/particles have been used extensively in surface assembly studies, particularly with alkanethiol monolayers because of their compatible chemistry ***(12)***.

3.2.1. Colloidal Silver

Two methods of preparing colloidal silver have been used extensively in the literature, which both involve the reduction of silver nitrate. The Lee and Meisel protocol involves the reductant citrate ***(13)*** whereas Blatchford and coworkers use borohydride ***(14)***. Citrate-reduced colloidal silver has advantages over its borohydride counterpart. A more uniform particle size distribution is produced, enabling the magnitude of scattering enhancement to be greater ***(15)***. The formation of an organic coating on the silver, during the citrate-reduction procedure, is important for biological applications. This organic layer, believed to be polymeric citrate ***(16)***, acts as a barrier preventing direct contact between the metal surface and adsorbed protein, minimizing protein denaturation. Borohydride-reduced Ag colloids lack an organic coating and therefore adsorbed proteins are more susceptible to denaturation ***(17)***.

The protocol for Lee and Meisel silver colloid (sol) are listed as follows.

1. It is vitally important to have ultra-clean glassware in order to prepare uniform, stable colloid. To rigorously clean the glassware and break down contaminants the strongly oxidizing mixture known as *aqua regia* is used. Addition of HNO_3 to HCl produces a deep red solution that forms slowly over about 30 min. During this time the vessel should not be stoppered otherwise pressure builds up and may result in a mini eruption, spraying highly caustic material within the immediate vicinity. Approximately 500 mL is prepared and placed into the 1 L glass round-bottomed reaction vessel together with the glass stirrer. Generally, soaking the glassware overnight in this medium is sufficient to remove all contaminants.
2. *Aqua regia* is decanted into its storage glass vessel. Both the glass stirrer and reaction flask are cleaned for a prolonged period of time by very gentle scrubbing with a warm soapy solution and nylon-bristled brush. This brush should only be

used for colloidal glassware and NOT other general cleaning uses in order to prevent possible contamination. Cleaning should progress for between 30 min and an hour, taking care not to scratch the inner surface of the flask. Both flask and stirrer should be thoroughly rinsed with distilled water.

3. Approximately 470 mL distilled water is placed into the reaction flask and the glass stirrer is attached to a motor. The reaction vessel is clamped to the motor stand and supported on a tripod. Modification of the stirrer's position within the flask ensures a smooth stirring action. The rate of stirring is such that the vortex generated almost reaches the bottom of the vessel. Rapid heating is performed with a bunsen burner.
4. At 50–60°C 90 mg of silver nitrate, dissolved in 30 mL distilled water, is added to the stirring solution. Continuous rapid heating and stirring is maintained until the *initial* signs of boiling. Note that poor and unstable colloids result when large and numerous "superheated" gaseous pockets are formed. Upon attaining initial boiling, 10 mL of 1% w/v day fresh trisodium citrate is immediately added and heating is reduced. Vigorous stirring was continued under this gentle boiling action for 75 min, to remove excess citrate. The final vol is adjusted to 500 mL with distilled water.
5. One to two min after the addition of citrate a pale yellow color develops which intensifies. Color changes occur over the next few minutes. The intense yellow color develops into an orange an then to an orange/"hint of green" solution. This green coloration is easily apparent at the top of the stirred suspension. The solution becomes greener as the reaction progresses. Throughout all these time-dependent color changes the solution/suspension should still retain clarity. A slight greying of the suspension (i.e., inability to see the rotating stirrer links or the center of solution) is consistent with the generation of larger particles. A final colloid with these characteristics will be unstable and have poorer scattering capabilities. These are undesirable properties and, with experience, one can tell whether a good or bad colloid has been generated after less than 30 min of preparation.
6. The absorption spectra of metal colloids are sensitive to the size and distribution of these discrete particles ***(18)***. Consequently, absorption studies are typically used as a qualitative measure of the goodness and dispersity of generated colloids. Silver colloidal suspensions with acceptable characteristics generally possess absorption maxima in the range ~400–415 nm with full widths at half height of 50–60 nm. It is known that citrate-reduced Ag colloid possessing these characteristics are almost monodispersed suspensions with particle diameters ~27–35 nm ***(16)***. Transmission electron micrographs have shown that these discrete particles are hexagonal/pentagonal in nature ***(19)***. Good colloids are stable for a minimum of 8–10 wk and should be stored in the dark.

Citrate-reduced colloids, which possess the desired characteristics, have traditionally been difficult to prepare and, consequently, many researchers in the SERRS field have used borohydride-reduced Ag colloids. The simpler protocol for this Ag sol, known as Creighton's colloid *(14)*, is detailed as follows.

1. Glassware should be rigorously cleaned and rinsed with distilled [or doubly distilled] water. Typically, beakers are used in this colloidal preparation. 300 mL of freshly prepared 2 m*M* borohydride solution is placed into an ice bath and allowed to reach equilibrium. This solution is stirred with a clean magnetic flea.
2. 100 mL of 2 m*M* silver nitrate is added dropwise to the ice-cold borohydride solution, which is being continuously stirred. The suspension produced is yellow in color.
3. Upon completion of the addition of silver nitrate, the solution is heated to remove excess borohydride and the final volume is made up to 500 mL with distilled water. The sol produced should have an absorption band around 400 nm.

3.2.2. Sample Preparation

1. Pure flavoprotein is thawed on ice and prepared in the desired buffer at the appropriate concentration and pH. Catalytic activity of the flavoprotein should be checked with biochemical assays such as those detailed in Chapters 5 and 11.
2. 1% w/v L-ascorbic acid is freshly prepared and a small volume (approx15 µL) is added to the buffered protein solution. The sample is incubated on ice for about 15 min in order to equilibrate. The reason for the addition of L-ascorbic acid is to protonate polar residues on the protein surface, thereby facilitating adsorption onto the negatively charged citrate-reduced silver colloid.
3. The incubated protein sample is added to precooled Ag sol. Addition of this sample should initiate aggregation of the colloid, which is left to equilibrate for 15–30 min. The sample is placed into a stoppered fluorescence cell or microtiter plate and maintained at the desired temperature. Volumes used in SERRS studies are generally between 400 µL–2 mL and protein concentrations 5–20 × 10^{-8} *M*.
4. Experiments are performed in exactly the same manner as outlined in **step 5.** in **Subheading 3.1.**

3.3. Semiquinone Formation

Semiquinone species can be generated using different methods and in the presence or absence of oxygen. These methods are outlined as follows.

1. Pure flavoprotein is slowly thawed by placing on ice. A buffer with the desired ionic strength and pH is prepared. This buffer contains the appropriate chemical(s) for the chosen reduction method employed. Protein activity should be measured using an applicable biochemical assay.
2. Both samples (protein and buffer) are placed in separate compartments in a Thunberg cell and oxygen can removed by repetitively vacuum pumping and backfilling the sample with oxygen-scrubbed argon or ultrahigh purity nitrogen. Solutions are mixed and can then be transferred to a capillary which is then sealed *(6)* or spectroscopically probed within the Thunberg cell.

3. As outlined in **step 1**, several methods are used for the reduction of flavoprotein. Mixing the flavoprotein with a natural reductant like reduced nicotinamide adenine dinucleotide phosphate (NADPH) for cytochrome-P450 reductase or L-lactate for flavocytochrome b_2 (FCB) or L-lactate monooxygenase (LMO) will result in the formation of relatively stable semiquinone species ***(6,20)***. On the other hand, artificial reduction methods employ light to achieve partial or full reduction of the flavoprotein (*see* Chapter 1).
4. Tegoni et al. ***(20)*** have used two light-induced methods to generate such species. One method involves the stirred mixture of degassed FCB or LMO (0.3–1.5 m*M*), EDTA (1–50 m*M*) and FMN (1.5–40 μM) illuminated by a UV filtered 500 W lamp for 10 min periods. Maximal generation of the semiquinone was achieved using at least three series of preillumination steps. The second method involves the laser beam (363.8 nm) as the photoreduction light source, with only EDTA (0.7–0.8 m*M*) and degassed protein present. By varying laser power, stirring rate and the focusing of the laser onto the sample variable amounts of semiquinone were produced.
5. Photo-reduced species are typically followed by UV/visible absorption spectroscopy on the sample. It should be noted that RR spectroscopy has also been used as a direct probe of semiquinone species within a sample ***(20)***.
6. Samples can now be probed spectroscopically. Generally, laser lines are used which are near in energy to an electronic transition of the semiquinone species generated. For example, NADPH-cytochrome P450 reductase forms neutral, blue semiquinones that are probed with yellow/orange/red laser lines such as 568.2 nm or 647.1 nm. On the other hand, FCB and LMO have product[pyruvate]-stabilized red, anionic semiquinones. These are studied using blue-shifted light (363.8 nm).

4. Interpretation

A typical spectrum is presented in **Fig. 1**. In order to make an interpretation of what the spectrum means for protein structure/function an understanding of the atomic basis of each vibrational mode is required. For guidance throughout this section the flavin isoalloxazine ring, and its atomic numbering, is depicted in **Fig. 2**.

Many groups have used theoretical methods, in conjunction with isotopic labeling studies, to accurately determine which atoms are involved in each vibrational mode and their relative energy contribution to that mode ***(23–25)***. **Table 1** summarizes the atomic characteristics associated with a few selected modes and reported frequency ranges.

As noted from **Table 1**, several bands are sensitive to hydrogen bonding at specific atoms on ring III of the flavin moiety. Tegoni et al. ***(20)*** noted an increase in the frequency of band III correlated with an increased hydrogen bonding environment around the N_1/N_5 for crystallographically-defined flavoproteins. This upshift is due to an increase in bond character at $C_{4a}C_{10a}$ as a result of hydrogen bonding at N_1/N_5. Band II is susceptible to hydrogen bond-

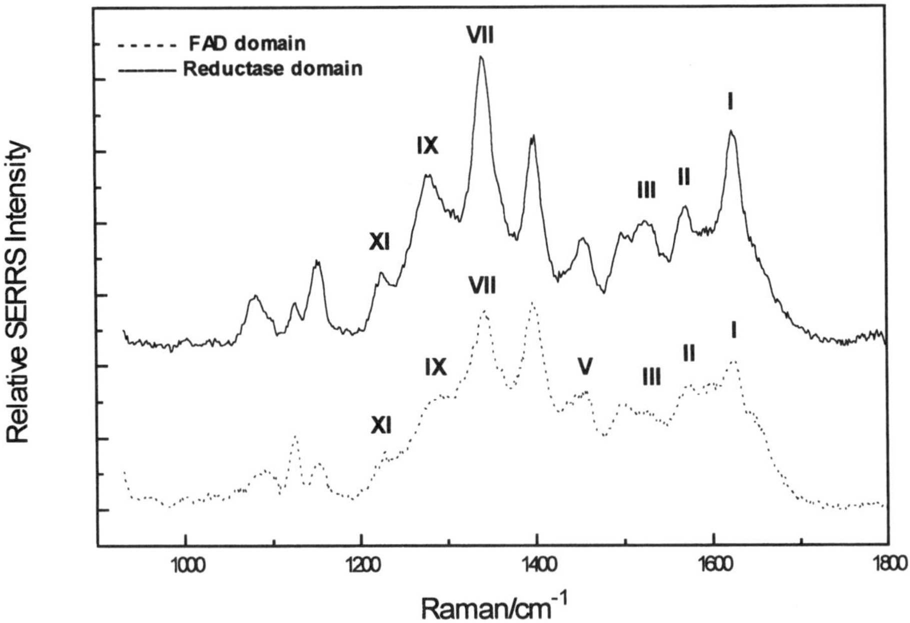

Fig. 1. Overlaid SERRS spectra of the reductase (1.7×10^{-7} *M*) and FAD/NADPH binding (2.0×10^{-7} *M*) domains of flavocytochrome P450 BM3 with excitation at 514.5 nm. Data reproduced with permission from ref. ***29***.

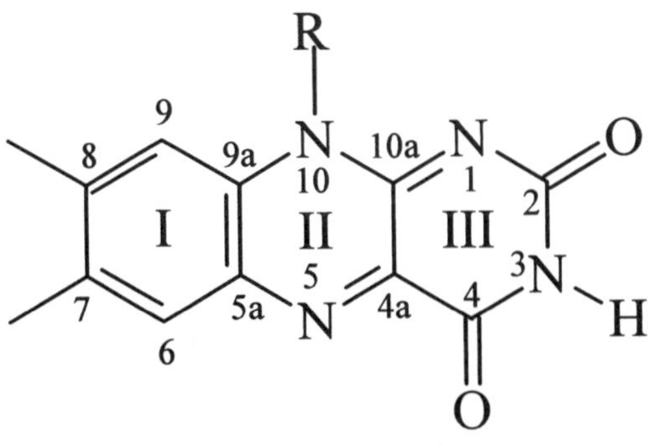

Fig. 2. The chemical structure of the isoalloxazine motif of flavins. The conventional ring numbering system, used throughout associated literature, is highlighted in the figure.

ing influences at $C_2 = O$ and $C_4 = O$ as well as N_1/N_5. The low frequency for band II upon adsorption of flavin onto citrate-reduced Ag colloid has been attributed to strong hydrogen bonding/electrostatic interactions between ring

Table 1
Selected Vibrational Modes for the Oxidized Isoalloxazine Moiety and Their Respective Frequency Range[a]

Vibrational nomenclature[a]	Atomic displacements[b,c]	Frequency range[d]
I	Ring I stretch	1624–1633
II	All bonds connected to $C_{4a}C_{10a}$	1571–1583
III	$\nu(C_{4a}C_{10a})$, $\nu(C_{4a}N_5)$, $\nu(C_{10a}N_1)$	1525–1559
VII	$\nu(C_{10a}N_{10})$, $\nu(C_{5a}C_{9a})$	1347–1356
IX	$\nu(C_4N_3)$ and methyl motions	1270–1288
X	$\nu(C_4N_3)$, $\nu(C_2N_3)$, $\delta(N_3H)$	1246–1263
XI	All bonds connected to N_{10} and $\nu(C_2N_3)$	1229–1235

All quoted frequencies are in cm^{-1}.
[a]Nomenclature adopted from ref. ***25***.
[b]Assignments from refs. ***23–25***.
[c]*See* **Fig. 2** for atomic numbering.
[d]Adapted from refs. ***21,23,26–29***.

II of the isoalloxazine moiety and the citrate coating on the Ag sol ***(29)***. Furthermore, the frequency of band X is thought to be an excellent marker of hydrogen bonding effects, arising from protein/solvent, at N_3-H ***(20,21,30)***. By analysis of crystallographically-resolved flavoprotein structures Tegoni ***(20)*** was able to identify a linear relationship between the frequency of band X and the number of hydrogen bonds at the positions N_1, O_2 and O_4, in addition to N_3-H. Using this relationship Macdonald et al. ***(29)*** have suggested up to six hydrogen bonds at these positions for the FAD moiety in an isolated FMN-deficient flavin domain of flavocytochrome P450 BM3. Kim and Carey ***(30)*** have noted a large upshift in frequency for a carbonyl vibration in riboflavin upon complexation with riboflavin binding domain. This large increase in hydrogen bonding at $C_4 = O$ is accompanied by a weakening of N_3-H bonding environment.

In addition to hydrogen bonding influences, the planarity of the isoalloxazine ring itself can be suggested by analysis of Raman frequencies. It has been discovered ***(20)*** that a strict linear relationship exists between the frequencies of bands II and III when the isoalloxazine ring is planar. No structurally characterized flavoproteins with planar flavin moieties deviate from this relationship. By analysis of these bands one can determine whether the prosthetic group of the flavoprotein studied is planar or not. Application of this relationship to cytochrome P450 BM3 has suggested that only upon the removal of the FMN binding domain does the FAD moiety become planar ***(29)***.

Within catalytic systems, hydrogen bonding and planarity are believed to be important in efficient and controlled delivery of electrons to/from flavin species. These structural factors are manifested in altered electrochemical redox couple(s) and Raman techniques have identified precise atomic perturbations which may be critical for catalysis. For example, an increase in bond length for $C_2 = O$ and $C_4 = O$ has been identified for the anionic semiquinones of FCB and LMO in the presence of its physiological product, pyruvate ***(20)***. Removal of the FMN-binding domain from flavocytochrome P450 BM3 changes both the hydrogen bonding environment around ring III and the planarity of the flavin moiety for this protein ***(29)***. It has been suggested by these authors ***(20,29)*** that the redox potential is very sensitive to polar/electrostatic effects on the $C_2 = O$, N_3-H, $C_4 = O$ structural hinge. This hinge, together with N_5, is also implicated as being involved in the catalytically important charge transfer substrate/flavoprotein complexes which occur in acyl CoA dehydrogenases and D-amino acid oxidase ***(31,32)***. It could be that this hinge is involved in catalytic events in many flavoproteins and Raman spectroscopies provide an excellent and sensitive probe of such events.

To conclude, Raman techniques are a powerful tool for elucidating structure/function relationships within flavoproteins, providing information on hydrogen bonding characteristics, ring planarity, and substrate/flavin complexes that aid in our understanding of the molecular mechanisms of these catalytic systems.

5. Notes

1. In almost all RR studies of flavins/flavoproteins fluorescence is a major interferent. To reduce fluorescence, quenching agents, such as KI, or spatial (remote) resolution, available to the technique of coherent antistokes Raman scattering (CARS) have been utilized for these systems ***(21)***. Variation of laser excitation line can also dramatically alter the quantum efficiency or extent of fluorescence. Whereas setting up a CARS experiment is unrealistic for many readers, simple addition of 2–3 *M* KI (or other quencher) or variation of laser line may be sufficient to reduce/eliminate fluorescent interference in their particular experiments.
2. Storage of *aqua regia* should be in glass vessels, loosely sealed with an inverted stopper. After two wk the intense red color will disappear and the solution becomes orange. An additional 2–4 wk results in a pale yellow color. It is desirable that colloidal glassware is cleaned using the "red" solution. Note that the orange/yellow solution of older *aqua regia* can still be used for cleaning glassware for other purposes.
3. If aggregation cannot be achieved by addition of the acidified protein solution to Ag colloid further acidification is required. This may take the form of additional L-ascorbic acid but can be done using different aggregating agents such as poly-L-lysine or spermine. These agents have been used to adsorb negatively charged species onto citrate-reduced Ag colloids ***(22)***. Lowering the pH of the buffering

solution is another possibility. It is best to avoid using inorganic acids like HCl/HNO_3 as stability and reproducibility of scattering efficiency are strongly perturbed. The likelihood of denatured protein is also greatly increased by the addition of these inorganic acids.

References

1. Spiro, T. G., ed. (1988) *Biological Applications of Raman Spectroscopy,* vols. 1–3, Wiley, New York, pp. 133–215, 439–490.
2. Unno, M., Christian, J. F., Benson, D. E., Gerber, N. C., Sligar, S. G., and Champion, P. M. (1997) Resonance Raman investigations of cytochrome $P450_{cam}$ complexed with putidaredoxin. *J. Am. Chem. Soc.* **119,** 6614–6620.
3. Hildebrandt, P., Hoffmann, A., Lindemann, P., Heibel, G., Braslavsky, S. E., Schaffner, K., and Schrader, B. (1992) Fourier transform resonance Raman spectroscopy of phytochrome. *Biochemistry* **31,** 7957–7962.
4. Mouro, C., Jung, C., Bondon, A., and Simonneaux, G. (1997) Comparative fourier transform infrared studies of the secondary structure and the CO heme ligand environment in cytochrome $P\text{-}450_{cam}$ and cytochrome $P\text{-}420_{cam}$. *Biochemistry* **36,** 8125–8134.
5. Chi, Z. H. and Asher, S. A. (1998) UV resonance Raman determination of protein acid denaturation—selective unfolding of helical segments of horse myoglobin. *Biochemistry* **37,** 2865–2872.
6. Sugiyama, Y., Nisimoto, Y., Mason, H. S., and Loehr, T. M. (1985) Flavins of NADPH-cytochrome P-450 reductase: evidence for structural alteration of flavins in their one-electron-reduced semiquinone states from resonance Raman spectroscopy. *Biochemistry* **24,** 3012–3019.
7. Horsburgh, M. J., Foster, T. J., Barth, P. T., and Coggins, J. R. (1996) Chorismate synthase from *Staphylococcus aureus. Microbiology* **142,** 2943–2950.
8. Rousseau, D. L., Friedman, J. M., and Williams, P. F. (1979) The resonance Raman effect, in *Topics In Current Physics,* vol. 11 (Weber, A., ed.), Springer-Verlag, Berlin, pp. 203–252.
9. Macdonald, I. D. G. and Smith, W. E. (1996) Orientation of cytochrome c adsorbed on a citrate-reduced silver colloid surface. *Langmuir* **12,** 706–713.
10. Creighton, J. A. (1988) The selection rules for surface-enhanced Raman spectroscopy, in *Spectroscopy of Surfaces,* vol. 16, (Clark, R. J. H. and Hester, R. E., eds.), Wiley, London and New York, pp. 37–89.
11. Cotton, T. M., Schultz, S. G., and Van Duyne, R. P. (1980) Surface-enhanced resonance Raman scattering from cytochrome c and myoglobin adsorbed on a silver electrode. *J. Am. Chem. Soc.* **102,** 7960–7962.
12. Meuse, C. W., Niaura, G., Lewis, M. L., and Plant, A. L. (1998) Assessing the molecular structure of alkanethiol monolayers in hybrid bilayer membranes with vibrational spectroscopies. *Langmuir* **14,** 1604–1611.
13. Lee, P. C. and Meisel, D. (1983) Surface-enhanced Raman scattering of colloid-stabilizer systems. *Chem. Phys. Lett.* **99,** 262–265.
14. Creighton, J. A., Blatchford, C. G., and Albrecht, M. G. (1979) Plasma resonance enhancement of Raman scattering by pyridine adsorbed on silver or gold sol par-

ticles of size comparable to the excitation wavelength. *J. Chem. Soc. Faraday Trans. II* **75,** 790–798.

15. Sheng, R.-S., Zhu, L., and Morris, M. D. (1986) Sedimentation classification of silver colloids for surface-enhanced Raman scattering. *Anal. Chem.* **58,** 1116–1121.
16. Munro, C. H., Smith, W. E., Garner, M., Clarkson, J., and White, P. C. (1995) Characterization of the surface of a citrate-reduced colloid optimized for use as a substrate for surface-enhanced resonance Raman scattering. *Langmuir* **11,** 3712–3720.
17. Smulevich, G. and Spiro, T. G. (1985) Surface enhanced Raman spectroscopic evidence that adsorption on silver particles can denature heme proteins. *J. Phys. Chem.* **89,** 5168–5173.
18. Mulvaney, P. (1996) Surface plasmon spectroscopy of nanosized metal particles. *Langmuir* **12,** 788–800.
19. Rodger, C., Smith, W. E., Dent, G., and Edmondson, M. (1996) Surface-enhanced resonance-Raman scattering—an informative probe of surface. *J. Chem. Soc. Dalton Trans.* **4,** 791–799.
20. Tegoni, M., Gervais, M., and Desbois, A. (1997) Resonance Raman study on the oxidized and anionic semiquinone forms of flavocytochrome b_2 and L-lactate monooxygenase. Influences of the structure and environment of the isoalloxazine ring on the flavin function. *Biochemistry* **36,** 8932–8946.
21. McFarland, J. T. (1987) Flavins, in *Biological Applications of Raman Spectroscopy Resonance Raman Spectra of Polyenes and Aromatics,* vol. 2 (Spiro, T. G., ed.), Wiley, New York, pp. 211–302.
22. Graham, D., Smith, W. E., Linacre, A. M. T., Munro, C. H., Watson, N. D., and White, P. C. (1997) Selective detection of deoxyribonucleic acid at ultra low concentrations by SERRS. *Anal. Chem.* **69,** 4703–4707.
23. Lively, C. R. and McFarland, J. T. (1990) Assignment and the effect of hydrogen bonding on the vibrational normal modes of flavins and flavoproteins. *J. Phys. Chem.* **94,** 3980–3994.
24. Abe, M., Kyogoku, Y., Kitagawa, T., Kawana, K., Ohishi, N., Takai-Suzuki, A., and Yagi, K. (1986) Infrared spectra and molecular association of lumiflavin and riboflavin derivatives. *Spectrochim. Acta* **42A,** 1059–1068.
25. Bowman, W. D. and Spiro, T. G. (1981) Normal mode analysis of lumiflavin and interpretation of resonance Raman spectra of flavoproteins. *Biochemistry* **20,** 3313–3318.
26. Lee, N-S., Hsieh, Y-Z, Morris, M. D., and Schopfer, L. M. (1987) Reinterpretation of surface-enhanced resonance Raman scattering of flavoproteins on silver colloids. *J. Am. Chem. Soc.* **109,** 1358–1363.
27. Holt, R. E. and Cotton, T. M. (1987) Free flavin interference in surface enhanced resonance Raman spectroscopy of glucose oxidase. *J. Am. Chem. Soc.* **109,** 1841–1845.
28. Cohen, J. D., Bao, W., Renganathan, V., Subramaniam, S. S., and Loehr, T. M. (1997) Resonance Raman spectroscopic studies of cellobiose dehydrogenase from *Phanerochaete chrysosporium. Arch. Biochem. Biophys.* **341,** 321–328.

29. Macdonald, I. D. G., Smith, W. E., and Munro, A. W. Analysis of the structure of the flavin-binding sites of flavocytochrome P450 BM3 using surface enhanced resonance Raman scattering. *Eur. Biophys. J.*, in press.
30. Kim, M. and Carey, P. R. (1993) Observation of a carbonyl feature for riboflavin bound to riboflavin-binding protein in the red-excited Raman spectrum. *J. Am. Chem . Soc.* **115,** 7015–7016.
31. Nashina, Y., Sato, K., Miura, R., and Shiga, K. (1995) Structures of charge-transfer complexes of flavoenzyme D-amino acid oxidase: A study by resonance Raman spectroscopy and extended Huckel molecular orbital method. *J. Biochem.* **118,** 614–620.
32. Nishina, Y., Sato, K., Hazekawa, I., and Shiga, K. (1995) Structural modulation of 2-enoyl-CoA bound to reduced Acyl-CoA dehydrogenases: a resonance Raman study of a catalytic intermediate. *J. Biochem.* **118,** 800–808.

10

NMR of Flavoproteins

Jacques Vervoort and Marco Hefti

1. Introduction

Flavoproteins are involved in a wide variety of enzymatic reactions like oxidation, reduction and mono-oxygenation. Consequently, flavin cofactors occur in many different forms including, for example, the oxidized semiquinone and reduced states as well as the C4a-(hydro)peroxyflavin form (**Fig. 1** and **Fig. 2**). In flavoprotein research the geometrical and electronic properties of all these different flavin species are of interest. This chapter deals with the use of nuclear magnetic resonance (NMR), a technique enabling the determination of detailed structural and electronic characteristics of flavoproteins and their cofactors.

2. Materials

2.1. Sample Preparation

NMR has often been rated as a nonsensitive technique, requiring samples of 0.25–1.5 mL vol at protein concentrations in the order of 1–4 m*M*. Thus, for a general NMR measurement on a flavin-containing protein of 60 kDa molecular mass, between 15–300 mg of protein is required. The precise amount of protein needed depends on the type of measurement, the type of NMR instrument, the temperature used, the solubility of the protein as well as the stability of the protein (NMR spectra need to be averaged over long periods of time). The upper limit for the molecular mass of the protein to be studied is about 200 kDa *(1)* because of limitations in solubility, increase in the number of NMR resonances with the increasing number of amino acids and the increase in linewidths of the NMR resonances due to smaller tumbling rates *(2,3)*. Altogether, NMR is only of interest for proteins that can be obtained in relatively

From: Methods in Molecular Biology, vol. 131: Flavoprotein Protocols
Edited by: S. K. Chapman and G. A. Reid © Humana Press Inc., Totowa, NJ

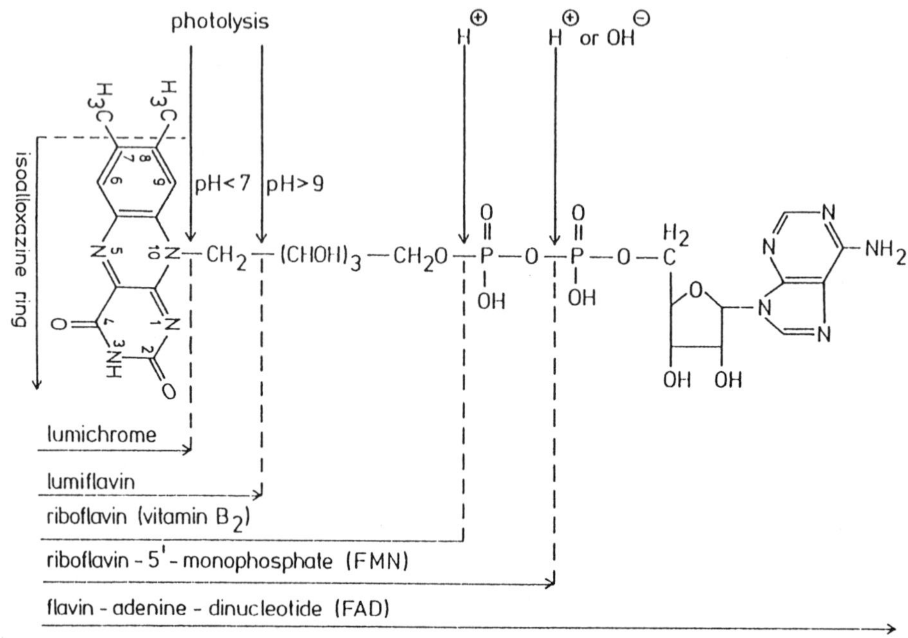

Fig. 1. The most common flavin cofactors.

large amounts, for example, in those cases where the corresponding genes have been cloned and overexpressed. The latter is also an important advantage for situations where the measurements require the preparation of ^{2}H, ^{13}C or ^{15}N enriched protein samples for performing NMR experiments including the specific advantages/possibilities provided by these nuclei as compared to the generally used ^{1}H NMR measurements. To this end genetically engineered organisms can be grown in media containing ^{15}N and/or ^{13}C labeled amino acids, which are commercially available (Isotec Inc., Miamisburg, OH; Martek Biosciences Corp., Columbia, MS), thus providing the possibility to produce specifically labeled proteins. The major advantage of flavoproteins is that specific enrichment of the cofactor can be performed synthetically or biosynthetically ***(4,5)*** and the flavin cofactor can be replaced with the ^{13}C or ^{15}N labeled derivative ***(6)***.

Furthermore, during sample preparation specific care should be taken to avoid the presence of paramagnetic ions/species, because the presence of unpaired electron spins causes significant broadening of the signals resulting in loss of resolution, especially in ^{31}P NMR. Only in a few cases, which are beyond the scope of this chapter, the presence of a paramagnetic center might be used to obtain specific structural information and NMR constraints using

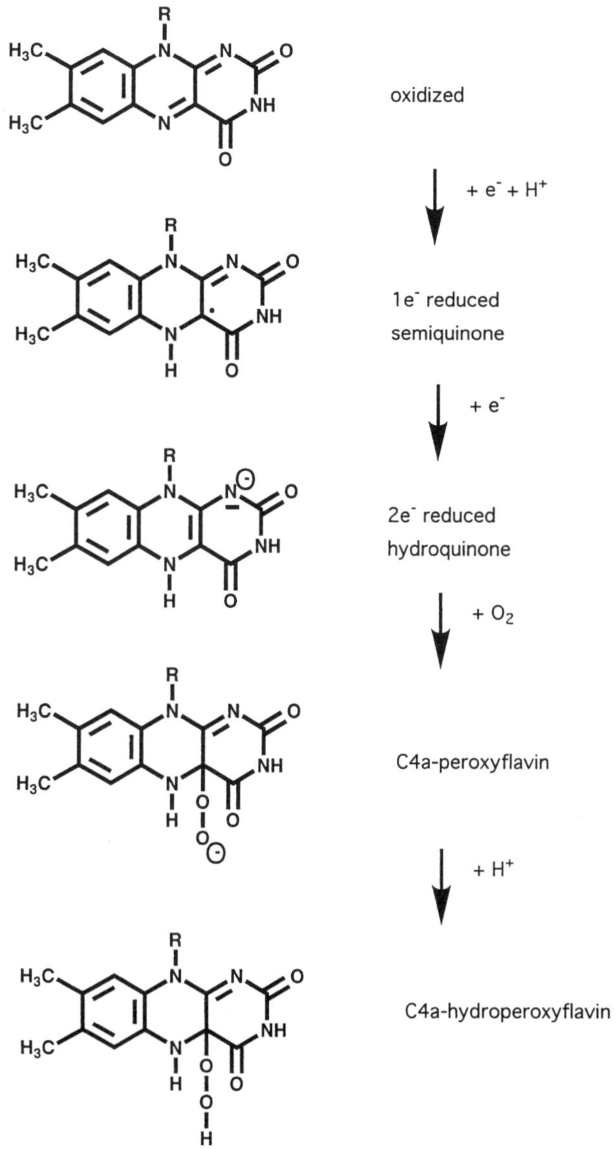

Fig. 2. Various forms of the flavin cofactor relevant for flavin-based biochemical reactions.

the large chemical shift values of the nuclei close to this center. For the use and limitations of NMR on paramagnetic centers the reader is referred to papers of the group of Bertini *(7)*.

Finally, as NMR measurements are usually performed at relatively high temperature and high protein concentrations, the solubility, but especially the stability of the protein sample should be investigated and checked under conditions as required for the measurement.

2.2. Instrumentation

Depending on the type of measurements and information required the instrument to be used can be defined. For initial measurements the use of a narrow-bore 9.4-T (400 MHz for ^{1}H) or 11.7-T (500 MHz) NMR instrument is sufficient. In principle these type of instruments are best suited for ^{13}C or ^{15}N NMR measurements of specifically labeled flavoproteins. However, for multidimensional NMR measurements it is often necessary in later stages of research, in order to solve the more crowded regions of a spectrum and/or to solve specific questions, to use a 14.1-T, 17.6-T MHz, or even a 18.8-T NMR (800 MHz for ^{1}H) instrument. Generally, these expensive NMR spectrometers are embedded within large scale facilities where one may apply for instrument time.

3. Method

In order to perform an NMR measurement on a flavin-containing protein, it is of importance to clearly define the type of information required, because this determines the type of NMR measurement to be performed. In many cases of interest for biochemists, research questions relate to structural information. To this end various multidimensional NMR measurements can be of use. The most important and commonly used three-dimensional NMR experiments are defined in **Table 1**. This overview, presented in **Table 1**, provides some insight in the type of experiments that can be performed, but in practice, it is advisable to also discuss the specific problem to be solved with a specialist in the field, who will also be the one actually performing the measurement. Given the problems encountered in using three-dimensional NMR for obtaining the desired structural information on proteins of increasing size, the main focus for flavoproteins has been on flavodoxins (M_r between 14–22 kDa). However, due to improvement in pulse sequences and instrumentation proteins up to 40 kDa can now be studied using three- or even four-dimensional NMR methods *(2)*. These improvements will make it possible to perform structural studies on flavoproteins other than flavodoxins. In spite of this, given the size of flavoproteins in general, most of them will remain beyond reach for three-dimensional structural NMR studies.

In contrast to three-dimensional NMR experiments, one-dimensional NMR experiments can be performed more easily and can already provide detailed information about the electronic and structural characteristics of the enriched flavin cofactor *(5,10)*. Using one-dimensional NMR ^{13}C or ^{15}N labeled fla-

Table 1
Three-Dimensional NMR Experiments Using a^{15}N or ^{13}C/^{15}N Labeled Protein Sample *(8,9)*

Experiment	Nucleus t_1	t_2	t_3
^{15}N-NOESY-HMQC	^{1}H	^{15}N	^{1}H
^{15}N-TOCSY-HMQC	^{1}H	^{15}N	^{1}H
HMQC-NOESY-HMQC	15N	^{15}N	^{1}H
HNHA	^{1}H	^{15}N	^{1}H
CBCANH	$^{13}C^{\alpha\beta}$	^{15}N	^{1}H
CBCA(CO)NH	$^{13}C^{\alpha\beta}$	^{15}N	^{1}H
HNCO	^{13}CO	^{15}N	^{1}H
HNCA	$^{13}C^{\alpha}$	^{15}N	^{1}H
CBCANH	$^{13}C^{\alpha}$	^{15}N	^{1}H
HBHA(CO)NH	$^{1}H^{\alpha\beta}$	^{15}N	^{1}H
C(CO)NH	$^{13}C^{\alpha\beta}$	^{15}N	^{1}H
HCCH-TOCSY	^{1}H	$^{13}C^{\alpha\beta}$	^{1}H

voproteins up to 200 kDa can be studied *(1)*. In pioneering studies performed by Moonen et al. *(11,12)* it was shown that the observed ^{13}C and ^{15}N chemical shift values of the isoalloxazine moiety of the flavin cofactor can be interpreted with a limited set of parameters. This finding made it possible, using specifically ^{13}C and/or ^{15}N labeled flavin cofactors bound to the protein studied, to interpret differences in the measured ^{13}C and ^{15}N chemical shift values to differences in electronic and structural configurations of the flavin cofactor *(5,13–15)*. Using ^{13}C NMR it was e.g., feasible to study the structural characteristics of the stable C(4a)-hydroperoxyflavin intermediate in bacterial luciferase *(10)*. This study not only settled a long dispute but it also provided essential structural details of the peroxyflavin intermediate. One of the indisputable outcomes of the studies on the electronic characteristics of the flavin cofactor was that in virtually all cases of the flavoproteins studied the flavin ring system was anionic in the two-electron reduced state, carrying a negative charge at the N(1) position *(15)*. Without the NMR studies performed this observation would have been practically impossible.

The nucleus most commonly used for NMR studies on most proteins is ^{1}H. The ^{1}H nucleus has almost 100% natural abundance and is present in all the amino acids of the protein, including their side-chains, the flavin cofactor as well as the water molecules. This implies that ^{1}H NMR measurements generally can provide information on all residues present in the structure. Thus ^{1}H NMR measurements can reveal information on, for example:

1. The position of solvent exchangeable hydrogen atoms providing insight into the solvent accessibility of parts of the protein structure.
2. The chemical shift values of individual hydrogen atoms, providing information on the electron density/hydrophobicity of their chemical surroundings.
3. The three-dimensional position of hydrogen atoms with respect to one another.
4. The relaxation behavior of the atoms providing information on, for example, their distance with respect to a paramagnetic center.

However, due to the strong increase in the number of ^{1}H resonances and problems of peak overlap with increasing size of the protein studied, the use of ^{1}H NMR for flavoproteins is limited to those proteins which are smaller than 40 kDa.

For specific purposes, information on chemical shift values, relaxation behavior, or NOEs of not only the hydrogen atoms, but also the carbon and/or nitrogen atoms present in the protein structure, can be of use. This provides information on the electron density distribution within the flavin cofactor, because the chemical shift of the N and C atoms is dependent on the electron density around the nucleus. The possibility to use ^{13}C and ^{15}N NMR is dependent on possibilities to obtain specifically ^{13}C or ^{15}N labeled proteins, because the natural abundance of these isotopes is 1.1%–0.4%, respectively. Using an enrichment of 90% or above (for ^{13}C or ^{15}N), the protein concentration needed for a one-dimensional NMR ^{13}C or ^{15}N experiment (overnight) on a 11.7-T or 14.1-T instrument is at least 2 m*M* (1.5 mL) but preferably 3–4 m*M* should be used. Improvements in instrumentation, especially using cryoprobes with superconducting coils at liquid nitrogen temperature or perhaps even room temperature, will give an increase in signal-to-noise of a factor of 2 to almost 4.

^{31}P NMR studies are quite easy to perform as the natural abundance of the ^{31}P isotope is 100% and the intrinsic sensitivity of ^{31}P NMR is relatively high. As a result numerous ^{31}P NMR studies have been done for different flavoproteins (the flavin cofactor has either one (FMN) or two (FAD) phosphate groups) (**Fig. 1**), but unfortunately the interpretation of the ^{31}P NMR data in structural terms has been less straight forward (*see* **ref. 15** and references therein).

4. Notes

In spite of its relatively low sensitivity, high protein requirements and molecular weight limits around 200 kDa, NMR, together with X-ray crystallography, is among the most versatile and powerful techniques for obtaining experimental data on structural and electronic information on flavoproteins. The type of pulse sequences and nature of NMR experiments that can be performed are many, but, when applying NMR to flavoprotein research there are some general pitfalls, as well as some pitfalls more specific for flavoproteins as such. The following sections discuss such problems or faults to be aware of, and also some ways on how to solve and/or overcome these problems.

4.1. Tubes to Be Used

Although apparently trivial, one should consider the type of NMR tubes best to use. To achieve the best possible resolution and minimize peak overlap field homogeneity within the sample is important and thus, one should use a tube of homogeneous thickness, providing minimal disturbance of the magnetic field. However, one should keep in mind that such tubes are fragile and may include the risk of sample loss due to breaking of the tube. Thus, choosing the tube will include finding a balance between resolution and risk on tube damage. Furthermore, it is of importance to use the same tube for a series of experiments to be compared. This seems trivial when comparing different types of NMR measurements on the same sample, but is also of importance when comparing the protein in, for example, its reduced or oxidized state.

It is our experience that when using 5 mm tubes, the best quality tubes (e.g., Wilmad 535-PP, Wilmad Glass, Buena, NJ) are preferable. In special cases, where precious samples are used or where a limited amount of sample is available, special 5 mm tubes, suitable for small volumes, are preferable: Shigemi microcell NMR tubes (Shigemi Inc., Allison Park, PA). These tubes have the advantage that magnetic susceptibility effects are neglectable even when using a very small volume (250 μL). The main disadvantage of these Shigemi tubes is that they are quite expensive, but compared to the rest of the costs involved, sample and instrumentation, these tubes are a good investment.

4.2. Shimming/Field Homogeneity/Peak Overlap

Field homogeneity is required to obtain sharp resonances, and, thus, optimal resolution and signal-to-noise. Taking into account the relatively narrow ppm range of ^{1}H NMR chemical shift values in diamagnetic proteins, including the flavoproteins, from –2 to +12 ppm, the risk and problems of peak overlap are serious, especially as the size of the protein increases ***(1,2)***. It is essential to undertake careful shimming of the sample to optimize field homogeneity before the measurement is actually started. Taking into account that the acquisition of one FID generally takes between 0.5–2 s, but also that the signal-to-noise ratio improves proportionally with the square root of the number of scans accumulated, it can be easily understood that it is worth spending an hour or more on shimming before starting the measurement. Remember that in case you do not use the best NMR tube available, shimming can be very difficult or even impossible at high field instruments. Recently the introduction of gradient shimming has been a major improvement in the ease with which the spectroscopist can shim his or her sample.

4.3. Paramagnetic Effects

Within the field of NMR on flavoproteins problems originating from paramagnetic contributions are twofold. They relate either to the presence of paramagnetic transition metal ions in the sample solvent, or to the intrinsic presence of a paramagnetic center in the protein under investigation. The effects originating from the presence of such a paramagnetic center in the solvent can disturb the measurements considerably, but, in some specific cases they may be used to obtain specific information as well. As one example the presence of Mn^{2+} ions in solution leads to severe broadening of ^{31}P resonances of the phosphate groups in FMN/FAD but this effect is only observed when the phosphate groups actually forms a complex with the Mn^{2+} ion. This aspect of line broadening on complex formation can be used to determine if the phosphate groups of the protein-bound flavin cofactor are solvent accessible or not.

Generally, the presence of paramagnetic ions in the solution of the protein should be avoided and the use of metal chelating agents during the purification procedure is advisable. However, because EDTA and many other metal chelating agents contain many ^1H resonances themselves, their use in preparation of samples for NMR studies is hampered and limited. Instead one may use a Chelex column (Bio-Rad, Hercules, CA) in order to eliminate the solubilized paramagnetic ions present in the sample.

Clearly the presence of paramagnetic ions within the protein itself cannot be avoided. Then one can only investigate if the influence of the paramagnetic center on the NMR characteristics can be used to obtain additional information *(7)*.

References

1. Crane, E. J., Vervoort, J., and Claiborne, A. (1997) ^{13}C NMR Analysis of the cysteine-sulfenic acid redox center of enterococcal NADH peroxidase. *Biochemistry* **36,** 8611–8618.
2. Clore, G. M. and Gronenborn, A. M. (1998) New methods of structure refinement for macromolecular structure determination by NMR. *Proc. Natl. Acad. Sci. USA* **95,** 5891–5998.
3. Evans, J. (1995) *Biomolecular NMR Spectroscopy.* Oxford University Press, Oxford, UK, pp. 1–54.
4. Spencer, R., Fisher, J., and Walsh, C. (1976) Preparation, characterization, and chemical properties of the flavin coenzyme analogues 5-deazariboflavin, 5-deazariboflavin 5'-phosphate, and 5-deazariboflavin 5'-diphosphate, 5'-5'-adenosine ester. *Biochemistry* **15,** 1043–1053.
5. Vervoort, J., Müller, F., O'Kane, D. J., Lee, J., and Bacher, A. (1986) Bacterial luciferase, a carbon-13, nitrogen-15 and phosphorus-31 nuclear magnetic resonance investigation. *Biochemistry* **25,** 8067–8075.
6. Van Berkel, W. J., Van den Berg, W. A., and Müller, F.(1988) Large-scale preparation and reconstitution of apo-flavoproteins with special reference to butyryl-

CoA dehydrogenase from Megasphaera elsdenii. Hydrophobic-interaction chromatography. *Eur. J .Biochem.* **178,** 197–207.

7. Banci, L., Bertini, I., and Luchinat, C. (1994) Two-dimensional nuclear magnetic resonance spectra of paramagnetic systems, in *Methods in Enzymology,* vol 239 (James, Th. L. and Oppenheimer, N. J., ed.), Academic Press San Diego CA, pp. 485–515.
8. Peelen, S., Wijmenga, S. S., Erbel, P. J. A., Robson, R. L., Eady, R. R., and Vervoort, J. (1996) Possible role of a short extra loop of the long chain flavodoxin from *Azotobacter chroococcum* in electron transfer to nitrogenase: completer ^{1}H, ^{15}N and ^{13}C backbone assignments and secondary solution structure of the flavodoxin. *J. Biomol. NMR* **7,** 315–330.
9. Garrett, D. S., Seok, Y.-J., Liao, D.-I., Peterkofsky, A., Gronenborn, A. M. and Clore, G. M. (1997) Solution structure of the 30 kDa *N*-terminal domain of enzyme I of the *Escherichia coli* phosphoenolpyruvate: sugar phosphotransferase system by multidimensional NMR. *Biochemistry* **36,** 2517–2530.
10. Vervoort, J., Müller, F., Lee, J., van den Berg, W. A. M., and Moonen, C. T. W. (1986). Identification of the true carbon-13 nmr spectrum of the stable intermediate II in bacterial luciferase *Biochemistry* **25,** 8062–8067.
11. Moonen, C. T. W., Vervoort, J., and Müller, F. (1983) A reinvestigation of the structure of oxidized and reduced flavin: carbon-13 and nitrogen-15 NMR study *Biochemistry* **23,** 4859–4868.
12. Moonen, C. T. W., Vervoort, J., and Müller, F. (1983) A carbon-13 NMR study on the dynamics of the conformation of reduced flavin *Biochemistry* **23,** 4869–4875.
13. Beinert, W.-D., Rüterjans, H., and Müller, F. (1986) Nuclear magnetic resonance studies of the Old Yellow Enzyme 2. ^{13}C NMR of the enzyme recombined with ^{13}C-labeled flavin mononucleotides *Eur. J. Biochem.* **152,** 581–590.
14. Vervoort, J., van Berkel, W. J. H., Müller, F., and Moonen, C. T. W. (1991) NMR studies of *p*-hydroxybenzoate hydroxylase from *Pseudomonas fluorescens* and salicylate hydroxylase from *Pseudomonas putida. Eur. J. Biochem.* **200,** 731–738.
15. Müller, F. (1992) *Chemistry and Biochemistry of Flavoenzymes,* vol. III, CRC Press Boca Raton FL, pp. 557–595.

11

Flavoprotein Resolution and Reconstitution

Florence Lederer, Heinz Rüterjans, and Garrit Fleischmann

1. Introduction

Flavins (FAD and FMN) are generally tightly bound to proteins; they are prosthetic groups and not cofactors, which implies that their dissociation constants are usually fairly small, between 10^{-7} and 10^{-10} *M*. It is therefore impossible in most cases to remove the flavin by simple dialysis. It is however useful to be able to resolve the flavin from a protein and be able to reconstitute the latter, either with the normal prosthetic group, so as to study the flavin binding parameters, or with flavin analogues, for structural and mechanistic purposes ***(1,2)***. In addition, in recent years, the use of isotopically labeled flavins has enabled the examination of flavin-protein interactions by (NMR) spectroscopy ***(3)*** and led to the need for large-scale procedures.

A number of protocols and variants thereof have been proposed over the years for removing flavins from proteins, since each protein has its own characteristics. These protocols have been reviewed by Husain and Massey ***(4)*** and Müller and van Berkel ***(5)***. Precipitation with ammonium sulfate at acid pH is one of the classical methods, with a variant that involves the simultaneous presence of 1 *M* KBr during precipitation. Alternatively, the protein can be dialysed against 1–2 *M* KBr at neutral or acidic pH, depending upon the protein. Other methods have been used more rarely: dialysis against low concentrations of denaturing agents such as guanidinium chloride, or addition of a high molarity of $CaCl_2$. More recently, a novel variant was proposed, in which the protein is kept adsorbed on a hydrophobic column in the presence of about 1 *M* ammonium sulfate and 1 *M* KBr. Lowering the pH of the buffer releases the flavin, which is eluted. The apoprotein can then be reconstituted either on the column or after elution ***(6)***. The reader is referred to the above-mentioned references for details and references.

From: Methods in Molecular Biology, vol. 131: Flavoprotein Protocols
Edited by: S. K. Chapman and G. A. Reid © Humana Press Inc., Totowa, NJ

This chapter describes the preparation of a deflavocytochrome b_2 which can be reconstituted in high yield, using a modified version of the immobilized protein protocol **(6)**. Flavocytochrome b_2 is usually classified among the complex flavoproteins, because each of its four subunits carries one FMN and one protoheme IX. Its three-dimensional structure, however, indicates a structural independence of its flavodehydrogenase and heme domains **(7)**. The former is a homologue of several FMN-dependent enzymes that all oxidize L-2-hydroxy acids and present the triose-phosphate-isomerase (TIM) barrelfold; the latter is a member of the family of b_5-like cytochromes **(8)**. For oxidized flavocytochrome b_2, flavin association and dissociation rates were found to be $1 \times 10^6\ M^{-1}\ s^{-1}$ and $2 \times 10^{-4}\ s^{-1}$, respectively, with a K_d of 0.2–0.5 n*M* **(9)**. The dissociation constant is lower for the reduced enzyme than for the oxidized one; furthermore, it was observed that simply freezing and thawing the enzyme raised the dissociation constant by a factor of 10^2–10^3 **(9)**. In addition, the K_d for the deflavoenzyme was found to be higher than for the native enzyme: 10^{-7} to 10^{-8} *M* **(9–11)**.

Several methods have been used in the past for resolving the flavin from the protein: the classical ammonium sulfate precipitation at pH 2, which leaves the heme domain intact **(12)**, or denaturation in the presence of guanidinium chloride, followed by renaturation in the presence of heme only **(13)**. Neither of these methods, however, achieved a high reconstitution yield. Subsequently, filtration on a molecular sieve column at acid pH, with immediate neutralization of the eluted deflavoprotein, afforded the means of working on smaller protein amounts, but necessitated a careful adjustment of pH and flow rate, and could not be adapted to large samples **(11,14)**. The method proposed here has been successfully used for reconstituting several µmoles of flavocytochrome b_2 with a slight molar excess of isotopically labeled FMN for NMR studies **(15)** and can also be scaled down for small amounts of material.

2. Materials

2.1. Flavin Removal

The protocol is given here for the treatment of 1 µmole protein. *See* **Note 1** for significantly different amounts of material. The overall procedure differs from that proposed by van Berkel et al. **(6)** in that all the steps are carried out at 5°C, and neither KBr nor ethylene glycol is used.

1. Flavocytochrome b_2. The recombinant enzyme is purified as described **(16)**. It is best kept frozen at –80°C as a concentrated reduced solution (≥ 100 µ*M*) in the presence of 10 m*M* L- or DL-lactate. Its monomer concentration is determined using the heme spectrum (ε_{red} = 183 mM^{-1} cm^{-1} at 423 nm) **(17)**.
2. Standard buffer: 0.1 *M* NaH_2PO_4/K_2HPO_4 buffer, 1 m*M* EDTA, pH 7.
3. Enzymatic assay solution: 20 m*M* L-lactate, 2 m*M* ferricyanide in the standard buffer.

4. Equilibration buffer: 50 m*M* K_2HPO_4, 1 *M* $(NH_4)_2SO_4$, 1 m*M* EDTA, pH 7.0.
5. Resolution buffer: 50 m*M* K_2HPO_4, 1 *M* $(NH_4)_2SO_4$, 1 m*M* EDTA, pH 4.5 (*see* **Note 2**).
6. Elution buffer: 0.025 *M* NaH_2PO_4/K_2HPO_4 buffer, 1 m*M* EDTA, pH 7.0.
7. 1 *M* NaH_2PO_4/K_2HPO_4 buffer, pH 7.0.
8. Chromatography column: Phenyl Sepharose CL-4B (Pharmacia & Upjohn, Bridgewater, NJ) (2 cm diameter, height: 3 cm (*see* **Note 1**)), connected to a peristaltic pump and a UV detector.

2.2. Reconstitution

1. Standard buffer (*see* **Subheading 2.1., item 2**).
2. 1 *M* L-lactate in standard buffer.
3. 10 m*M* FMN in standard buffer, protected from light.
4. 10 m*M* sodium sulfite in standard buffer, freshly prepared.

3. Method

3.1. Flavin Removal

All the chromatographic steps should be carried out in the cold room, where the buffers will have been stored beforehand. The enzyme must be kept on ice outside the cold room.

1. Suspend the Phenyl Sepharose in the equilibration buffer, pour it into the column and run 3–4 column vol of equilibration buffer through it. Check the pH and the ionic strength of the effluent.
2 Thaw the flavocytochrome b_2 sample on ice. Determine its concentration if necessary (*see* **Subheading 2.1., step 1**). Check its specific activity; this is done at 30°C, by looking at ferricyanide reduction at 420 nm ($\varepsilon_{\text{ox-red}}$ = 1.04 mM^{-1} cm^{-1}). A good enzyme preparation should have a specific activity under these conditions of between 450–550 electrons transferred per second. Add solid ammonium sulfate to 1 *M* while stirring with a magnetic bar and maintaining the pH at 7 with 2–3 *M* KOH (145 mg ammonium sulfate added per mL enzyme solution)
3. Adsorb the sample onto the column at a flow rate of about 25 mL/h/cm^2 (*see* **Note 3**). Wash the column with equilibration buffer, at a flow rate of 90–100 mL/h/cm^2, until all unbound material has come out, as monitored at 280 nm. At this point, the color of the protein on the column has changed from salmon pink to red, indicating that the prosthetic groups have been oxidized by the oxygen present in the buffers (*see* **Note 4**). About 100 mL is sufficient for this step.
4. Elute the flavin by running the resolution buffer through the column at a flow rate of 90–100 mL/h/cm^2, until all the yellow FMN has come out; the completeness of elution is best monitored at 280 nm (*see* **Note 5**). About 150 mL is necessary.
5. Reequilibrate the column at neutral pH with equilibration buffer, at the same flow rate (about 100 mL). Monitoring the pH of the effluent helps limit the duration of this step.

6. Change to the elution buffer, at a flow rate of about 50 mL/h/cm². After about 2 column vol, the red protein starts coming out; elution is complete in about 2–3 column vol (*see* **Note 6**). The protein should be collected in test tubes containing a volume of 1 *M* phosphate buffer, pH 7, sufficient to bring back the fractions to a buffer concentration of about 0.1 *M* as soon as the protein has eluted (*see* **Note 7**).
7. Pool all the red fractions and determine the residual specific activity as described above. It should be around 1–3% that of the starting enzyme. The protein yield is on the order of 60–70%, based on the heme content (*see* **Note 8** for storage conditions of the deflavoenzyme).

3.2. Reconstitution

In principle, the protein can be reconstituted directly on the column after *see* **Subheading 3.1., step 5** above, by running buffer containing the desired analog through the sample before *see* **Subheading 3.1., step 6**. This procedure is, not recommended, however, because there is no check on the quality of the apoprotein preparation, and the procedure requires a larger amount of prosthetic group than reconstitution after elution. The protocol indicated below can be carried out immediately after elution and analytical determinations, or after storage at –80°C by thawing the deflavoenzyme on ice (*see* **Note 8**).

1. Add to the enzyme solution the required vol of 1 *M* lactate to bring the substrate concentration to 10 m*M*. Add the required vol of 10 m*M* flavin solution in standard buffer so as to have a slight excess of flavin over heme (*see* **Note 9**). Bring the solution to 30°C for 10 min, then place on ice again (*see* **Note 10**). At this point the heme has changed color to salmon pink, because it can be reduced by lactate via the incorporated flavin.
2. Determine the specific activity of the reconstituted enzyme as above, based on the heme content.
3. Determine the flavin content of the preparation by difference spectroscopy between sulfite-complexed enzyme and free enzyme ***(18)***. This has to be done with oxidized enzyme, but free flavin does not interfere. A 10- to 20-fold dilution of the reconstituted protein into lactate-free buffer may be sufficient for protein autoxidation, which can be followed spectrophotometrically (the oxidized heme Soret band has its maximum at 413 nm ***(17)***). An exactly identical vol of enzyme solution (between 5–10 μ*M* based on the heme concentration) is placed in two spectrophotometer cuvets. The base line is recorded. A minimal vol of sulfite (final concentration: 0.05–0.1 m*M*) is added to the reference cuvet and the same buffer vol added to the sample. For the sake of precision, the addition should be done without removing the cuvets from their holders. The difference spectrum is then recorded between 500–400 nm. The flavin concentration is calculated using $\varepsilon_{454} = 10.5\ mM^{-1}\ cm^{-1}$ ***(18)***. The reconstitution yield can then be calculated by comparing the flavin and heme concentrations; it normally lies between 70–90%.

The specific activity per flavin can be compared to that of the native protein; it normally lies between 90–100% (*see* **Note 11**).

4. Notes

1. The protocol can be used for smaller quantities (50–100 nmoles), as well as larger ones (several µmoles). The Phenyl Sepharose bed vol should be around 8–9 mL per µmole protein. The column height should be as small as possible, for a rapid elution. As the protein is colored, it can be visualized on the column; it suffices that a few mm of the bed at the bottom remain white. The overall yield is somewhat better with the larger than the smaller amounts.
2. The pH is adjusted at room temperature to 4.40, no higher. Cooling at 4°C raises it by 0.1 unit (*see* **Note 5**).
3. The original protocol included 1 *M* KBr in the equilibration buffer, and manipulations were carried out at room temperature ***(6)***. At 4°C, salts crystallize out from this highly concentrated buffer. In the absence of ammonium sulfate, up to 2 *M* KBr does not lead to enzyme adsorption onto the column. The protocol was tested once at room temperature; in the absence of KBr, the reconstitution yield was poor. In its presence, it was about the same as that obtained at 4°C in its absence. The enzyme is generally more stable at 4°C, and where no air conditioning is available, work at 4°C guarantees a better reproducibility.
4. The enzyme must be in the oxidized state, because the flavin is more tightly bound when reduced and is more difficult to elute.
5. Flavin elution was also tested at pH 4. Under these conditions, it is more rapid and the residual activity of the deflavo preparation is even lower than at pH 4.5. But then the specific activity per flavin after reconstitution is significantly lower than that of the native enzyme. pH 4.5 is the highest one compatible with complete flavin elution. This value is critical for the best resolution. When large samples will be handled, it is wise to use a small protein amount first to check that the elution buffer pH is acidic enough.
6. Enzyme elution with 50 m*M* phosphate buffer is slower and, therefore, reconstitution yields can be affected. The original protocol ***(6)*** used ethylene glycol in the elution buffer.
7. The holoenzyme loses activity at low ionic strength. The stability of the deflavoenzyme with respect to this factor has not been checked.
8. Upon standing, the deflavoenzyme slowly loses reconstitutability at 4°C, especially in dilute solutions (around 10 µ*M*). At higher concentrations (0.1 m*M* and above), it can be kept for at least 24 h, without losing reconstitutability. If it is desired to keep it for a longer time, it can be stored frozen at –80°C for several weeks, after losing up to 10% reconstitutability during the freezing process. It is often desirable to have a concentrated enzyme stock solution, especially as dilute solutions tend to be unstable. The deflavoenzyme can be concentrated before freezing and reconstitution, using various ultrafiltration devices depending on the volume to be concentrated (Amicon cells, centricon concentrators, etc).

9. The minimum molar excess is calculated on the flavin-5'-phosphate content of the prosthetic group preparation; if it is pure, a 1.1-fold excess is sufficient. Commercial FMN samples contain on the order of 20% impurities (riboflavin-4'-phosphate among others). These impurities do not interfere with reconstitution and the nominal molar ratio can be increased without inconvenience. If it is desired, the flavin preparation can be purified by ion exchange chromatography ***(19)***.
10. At 4°C, reconstitution is biphasic: after a rapid phase during which about two-thirds of the activity is recovered, a slow increase is noted and maximal activity has not been reached after one hour. Warming to 30°C, as advised in ***(4)***, accelerates the process.
11. Reconstitution yields are usually higher at high protein concentrations; it is therefore advisable to concentrate the deflavoenzyme before reconstitution.

References

1. Ghisla, S. and Massey, V. (1986) New flavins for old: artificial flavins as active site probes of flavoproteins. *Biochem. J.* **239,** 1–12.
2. Murthy Y. V. S. N. and Massey, V. (1997) Synthesis and applications of flavin analogs as active site probes of flavoproteins. *Methods Enzymol.* **280,** 436–460.
3. Müller, F. (1992) Nuclear magnetic resonance studies on flavoproteins, in *Chemistry and Biochemistry of Flavoenzymes,* vol. 3 (Ed. F. Müller, F., ed.), CRC Press, Boca Raton, FL, pp. 557–595.
4. Husain, M. and Massey, V. (1978) Reversible resolution and reconstitution of flavoproteins into apoproteins and free flavin. *Methods Enzymol.* **53,** 429–437.
5. Müller F. and van Berkel, W. J. H. (1991) Methods used to reversibly resolve flavoproteins into the constituents apoprotein and prosthetic group, in *Chemistry and Biochemistry of Flavoenzymes* (Müller, F., ed.), CRC Press, Boca Raton, FL, pp. 261–274.
6. van Berkel, W. J. H., van den Berg, W. A. M., and Müller F. (1988) Large-scale preparation and reconstitution of apo-flavoproteins with special reference to butyryl-CoA dehydrogenase from *Megasphaera elsdenii*. Hydrophobic-interaction chromatography. *Eur. J. Biochem.* **178,** 197–207.
7. Xia, Z. X. and Mathews, F. S. (1990) Molecular structure of flavocytochrome b_2 at 2.4 Å resolution. *J. Mol. Biol.* **212,** 837–863.
8. Lederer, F. (1991) Flavocytochrome b_2, in *Chemistry and Biochemistry of Flavoenzymes,* vol. 2 (Müller, F., ed.), CRC Press, Boca Raton, FL, pp. 143–242.
9. Iwatsubo, M. and Di Franco, A. (1965) Groupes prosthétiques de la lactic-odéshydrogénase de la levure. III. Etude fluorométrique de la cinétique et de l'équilibre de combinaison du FMN à la partie protéique. *Bull. Soc. Chim. Biol.* **47,** 891–910
10. Baudras, A. (1965) Groupes prosthétiques de la L-lacticodéshydrogénase de la levure. II. Etude enzymatique de l'équilibre d'association de la flavine à l'hémoprotéine. *Bull. Soc. Chim. Biol.* **47,** 1177–1201
11. Pompon, D. and Lederer, F. (1979) Deazaflavins as cofactors for flavocytochrome b_2 from baker's yeast. *Eur. J. Biochem.* **96,** 571–579.

12. Baudras, A. (1965) Groupes prosthétiques de la L-lacticodéshydrogénase de la levure. I. Mise en évidence du rôle de la flavine par l'étude des propriétés de l'apo-L- lacticodéshydrogénase. *Bull. Soc. Chim. Biol.* **47,** 1143–1175.
13. Mével-Ninio, M., Pajot, P., and Labeyrie, F. (1971) Reconstitution of cytochrome b_2 following prosthetic groups dissociation by guanidine hydrochloride. Protoheme binding. *Biochimie* **53,** 35–41.
14. Pompon, D. and Lederer, F. (1978) Binding of Cibacron blue F3GA to flavocytochrome b_2 from baker's yeast. *Eur. J. Biochem.* **90,** 563–569.
15. Rüterjans, H., Fleischmann, G., Knauf, M., Löhr, F., Blümel, M., Lederer, F., Mayhew, S., and Müller, F. (1996) NMR-studies of flavoproteins. *Biochem. Soc. Trans.* **24,** 116–121.
16. Dubois, J., Chapman, S. K., Mathews, F. S., Reid, G. A., and Lederer, F. (1990) Substitution of Tyr254 with Phe at the active site of flavocytochrome b_2: consequences on catalysis of lactate dehydrogenation. *Biochemistry* **29,** 6393–6400.
17. Labeyrie, F., Baudras, A., and Lederer, F. (1978) Flavocytochrome b_2 or L-lactate cytochrome *c* reductase from yeast. *Methods Enzymol.* **53,** 238–256.
18. Lederer, F. (1978) Sulfite binding to a flavodehydrogenase, cytochrome b_2 from baker's yeast. *Eur. J. Biochem.* **88,** 425–431.
19. van Schagen, C. G. and Müller, F (1981) A ^{13}C nuclear magnetic resonance study of free flavins and *Megasphaera elsdenii* and *Azotobacter vinelandii* flavodoxin. *Eur. J Biochem.* **120,** 33–39.

12

Flavoenzyme Structure and Function

Approaches Using Flavin Analogues

Dale Edmondson and Sandro Ghisla

1. Introduction

Flavoenzymes are redox proteins that catalyze a wide diversity of biological reactions ranging from O_2 activation, to aromatic hydroxylations, dehydrogenations, and reactions in which they can accept or donate one or two electrons. In addition, they can fulfill structural and regulatory roles. This diversity is due to the fine tuning of the reactivity of the isoalloxazine ring of the flavin coenzyme by specific interactions with the protein moiety on the particular enzyme it complexes with. Methods to provide specific information on these interactions have relied on three approaches:

1. Determination of the three-dimensional structure of the flavoenzyme by X-ray diffraction techniques.
2. Spectroscopic probes of the protein influence on flavin structure by techniques such as nuclear magnetic resonance (NMR), resonance Raman, fluorescence, electron paramagnetic resonance (EPR), and circular dichroism (CD) spectroscopies.
3. Replacement of the native flavin coenzyme with suitable flavin analogues designed to ask specific questions regarding the reactivity, the accessibility, and the mode of specific interactions with the protein.

The first approach is labor intensive, requires the enzyme to crystallize in a form that diffracts with high resolution, and requires a knowledge of the protein's amino acid sequence for interpretation of the electron density maps. The second approach requires the synthesis of isotopically-labeled flavins (for NMR, resonance Raman, and EPR spectroscopies) to facilitate interpretation

From: Methods in Molecular Biology, vol. 131: Flavoprotein Protocols
Edited by: S. K. Chapman and G. A. Reid © Humana Press Inc., Totowa, NJ

of spectral data. These techniques also require reasonably large quantities of purified flavoenzyme (especially for NMR analysis) and access to specialized instrumentation and to investigators knowledgeable in these techniques. The third approach provides a good deal of unique information on the environment and reactivity of the flavin in its binding site, does not require large quantities of enzyme, and does not necessarily require access to highly specialized instrumentation. The major requirements are that the flavoenzyme is amenable to reversible resolution of the flavin coenzyme (*see* Chapter 11) and reconstitution of the apoenzyme with suitable flavin coenzyme analogues. This approach does require the availability of the riboflavin form of these analogues, which, in general, have to be obtained by chemical synthesis. Furthermore, the riboflavin analogues have to be enzymatically converted to either coenzyme form (FMN or FAD). This is conveniently done enzymatically with the *Brevibacterium* FAD synthetase system *(1)*. Once the FAD analogue is prepared, it can readily be converted to the FMN level by treatment with nucleotide pyrophosphatase.

Here, we will outline the basic approaches which can be applied to the study of those flavoenzymes containing noncovalently bound flavin coenzymes. Those containing covalently-bound flavins have not been amenable to this approach for obvious reasons. Recent developments in the area of covalent incorporation of flavins into members of this class of flavoenzymes (e.g., *p*-methyl cresol hydroxylase *(2)* and monoamine oxidases *(3)*) should provide the technology for similar types of approaches in this class of flavoproteins.

2. Historical Perspective

Flavin coenzymes are comprised of the redox-active isoalloxazine ring, a ribityl side chain, and (for FMN) a 5'-terminal phosphate ester or (for FAD) a pyrophosphate linkage of FMN with an aminomonophosphate (AMP) moiety (**Fig. 1**). Each of these moieties is intimately involved in binding the coenzyme to its site on the apoenzyme. Until recently, the ribityl side chain and the level of phosphorylation was thought only to be involved in binding interactions and that all of the chemical reactivity of the flavin in catalysis resided only in the ring system. Although this view may be true for the vast majority of flavoenzymes, it is not universally valid with the recent discovery of the required participation of the 2' OH group of the ribityl side chain in acyl-CoA dehydrogenase catalysis *(4)*. However, the bulk of informative work with modified flavins has been performed with ring-modified analogues.

The synthetic procedures for synthesis of ring-modified riboflavin analogues were developed mainly in Lambooy's and Hemmerich's laboratories (**refs.** ***5–7*** for synthetic procedures). Approaches developed to monitor the reactivities of

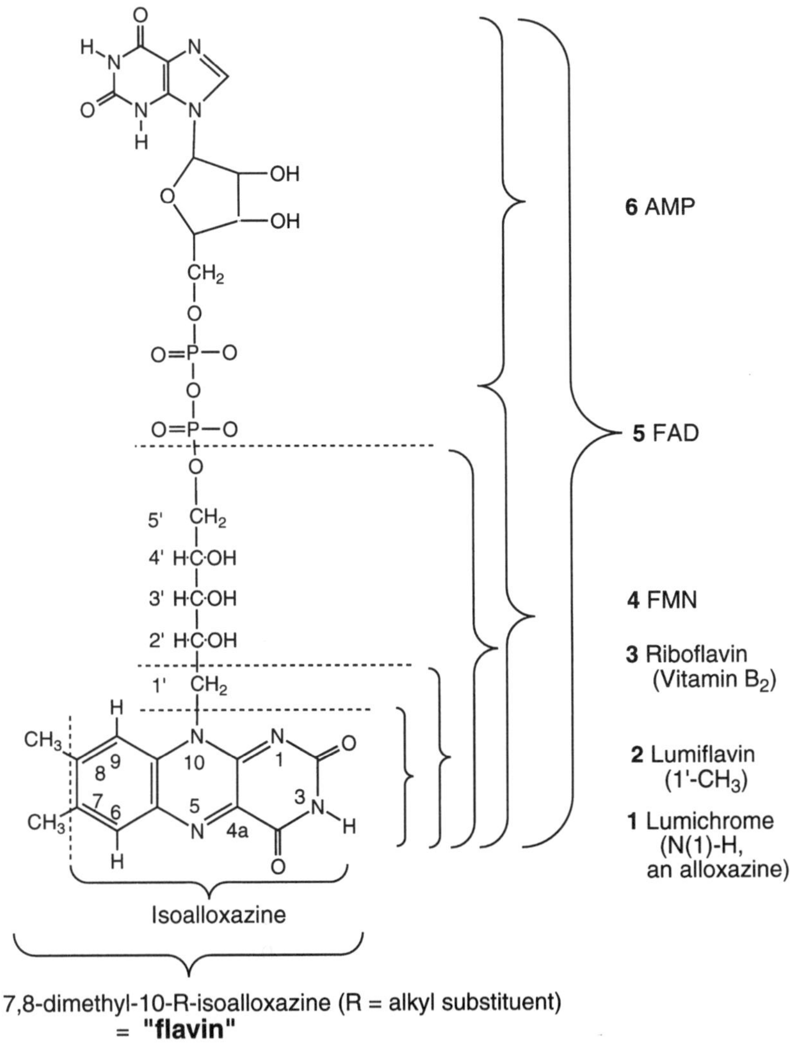

Fig. 1. Structures of various derivatives of the flavin. Note that the term "flavin" refers to an isoalloxazine molecule carrying two methyl groups at the positions 7, and 8, and a further substituent at position N(10). The ring numbering shown is that approved by the International union of Pure and Applied Chemistry. Different numbering systems are to be found in the older literature.

the flavin analogues when bound to the protein were developed in Massey's laboratory with the collaboration of one of the co-authors (S.G.). The enzymic conversion of riboflavin analogues to their FAD and FMN coenzyme forms is described in detail in *(8)*.

3. Rationale and Basic Considerations.

The information that can be obtained from studies with modified flavins can be subdivided into three distinct topics of study:

1. *Molecular environment of the isoalloxazine ring system, and accessibility* (study of the interaction of the flavin with its protein environment, specifically with amino acid groups, and of the regulation or prevention of access to flavin sites by reactive molecules.
2. *Mechanistic studies* (approaches aimed at the differentiation of catalytic mechanism reaction types).
3. *Modulation of the oxidation-reduction potential* (studies aimed at the understanding of the fine tuning of the flavin reactivity via its redox potential).

3.1. Flavin Interactions and Accessibility

As mentioned above, the ribityl side chain of the flavin serves in anchoring the catalyst to the protein. This chain, in both FAD and FMN, has been found from crystallographic studies on a number of flavoenzymes to be buried, at least partially, inside the protein. The interactions of the coenzyme side chain with the protein contribute the majority of the binding energy to the energetics of FMN or FAD binding to their respective apoenzymes. The 7,8-dimethylisoalloxazine moiety, i.e., the chromophoric part of the flavin, is an amphoteric molecule in that its xylene aromatic ring is lipophilic, while the pyrimidine subnucleus is hydrophilic. The discussion of accessibility/reactivity of the flavin can thus be subdivided into those of the pyrimidine ring and of the benzenoid domains (*see* **Fig. 2**) Already a superficial inspection of the structures (**Figs. 1** and **2**) make it clear that the pyrimidine ring will undergo specific interactions (i.e., hydrogen bonding) with protein functional groups. The mode of such interactions, as well as the accessibility of the pyrimidine moiety has been probed in various ways:

1. Position N(3)-H can be altered by alkylation (**Fig. 3**), and the size and properties of the N(3)-substituent can be varied. This approach will verify if the N(3)-H group undergoes an important interaction. 3-Methyl-FMN and FMN-3 acetate have been used for this purpose *(9)*. In general, most flavoproteins do not appear to be very tolerant of substantial changes at the isoalloxazine N(3) position.
2. The C(2)=O carbonyl, which is part of the pyrimidine amidine system, has been replaced with =N-R and =S. This opens two types of approaches. On the one hand the bulkier sulfur atom and its different chemistry compared to oxygen, can disturb the strength and geometry of H-bonds. These effects can to some extent balance their effects on the flavin ring in that the stronger electronegativity of the sulfur can compensate for the weaker H-bonds that it forms. The thiocarbonyl function, on the other hand, is a much better nucleophile and electrophile compared to the essentially unreactive C(2)=O group. Thus, 2-thioflavins will react

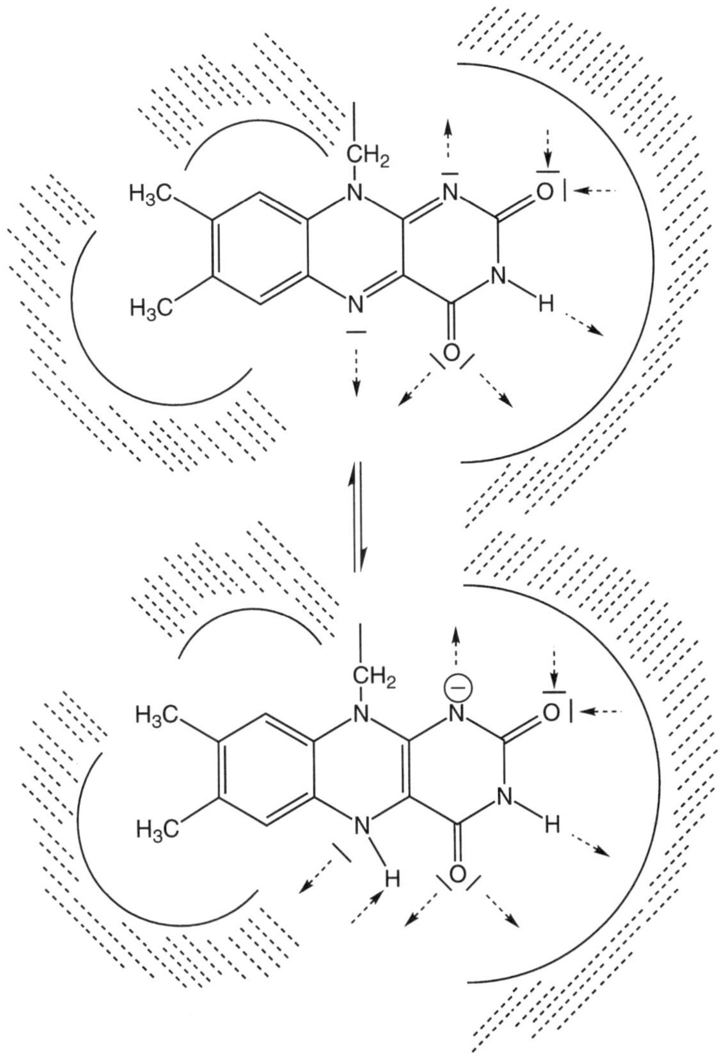

Fig. 2. Modes and positions of interaction of the various flavin functional groups with the protein. The latter is symbolized by the shaded areas. Arrows denote the possibility to form donor (→) or acceptor (←) hydrogen bonds. The open areas indicate positions at which accessibility has been found in a majority of cases. This does not necessarily apply to all flavoproteins.

with a variety of alkylating agents and are prone to oxidation ***(10)***. The (2)S group can also easily be substituted with amines and other nucleophiles, which are listed in **Fig. 4**. Upon determination of the basic reactivity of the (2)=S func-

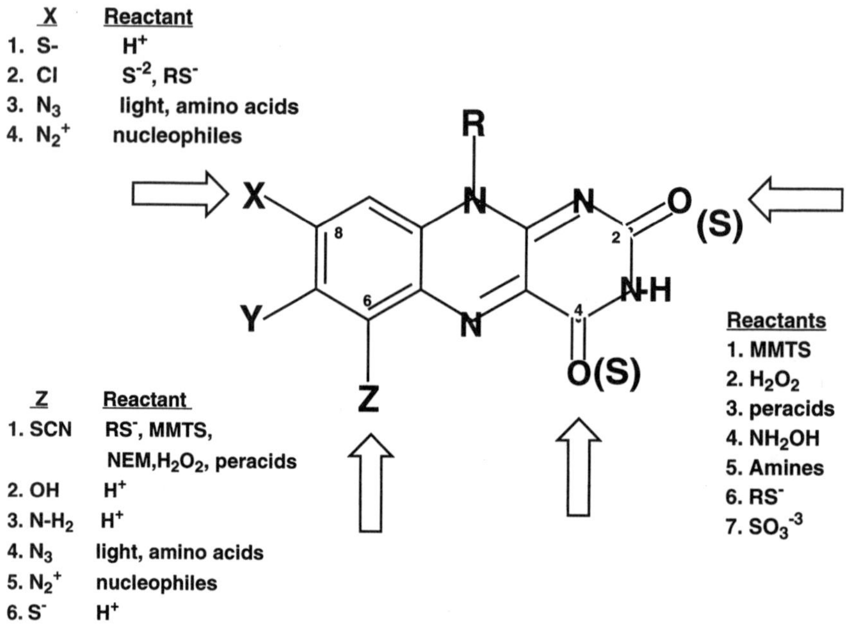

Fig. 3. Overview of the functional groups on the flavin ring that are amenable to chemical analysis by either modification or substitution. The arrows indicate the positions for which most of the studies have been done. This does not exclude that other reactive groups could be introduced in the flavin ring system. Study of the reactivity with the reactants shown can be used to monitor either the accessibility of the specific flavin position when bound to a protein, or its (modified) reactivity.

tion with a given reactant, the rate of reaction of the enzyme bound (2)=S flavin will reflect the degree of accessibility of this position from the bulk solvent. The approach has been implemented (e.g., with *p*-hydroxybenzoate hydroxylase) reconstituted with 2-thio-FAD and H_2O_2 as reactant ***(11)***.

3. The same type of approach can be implemented at position C(4)=O. In general, it should be stated that modifications at these positions are more labile, than the similar C(2)=O position discussed above. Two publications deal with this topic *(10,12)*. One particularly useful property of 4-thioflavins should be mentioned. The differentiation between the neutral and the anionic forms of reduced flavins bound to various enzymes is, in general, very difficult due to the spectral similarity of the two species, and the lack of particular features. In contrast, the reduced species of 4-thio-flavins exhibit well resolved absorption spectral bands with the maxima at 425 nm for the anionic and 485 for the neutral forms ***(10)***.

The (middle) pyrazine ring of the flavin system comprises the main locus of chemical reactivity, i.e., the positions C(4a)-N(5), which are involved in elec-

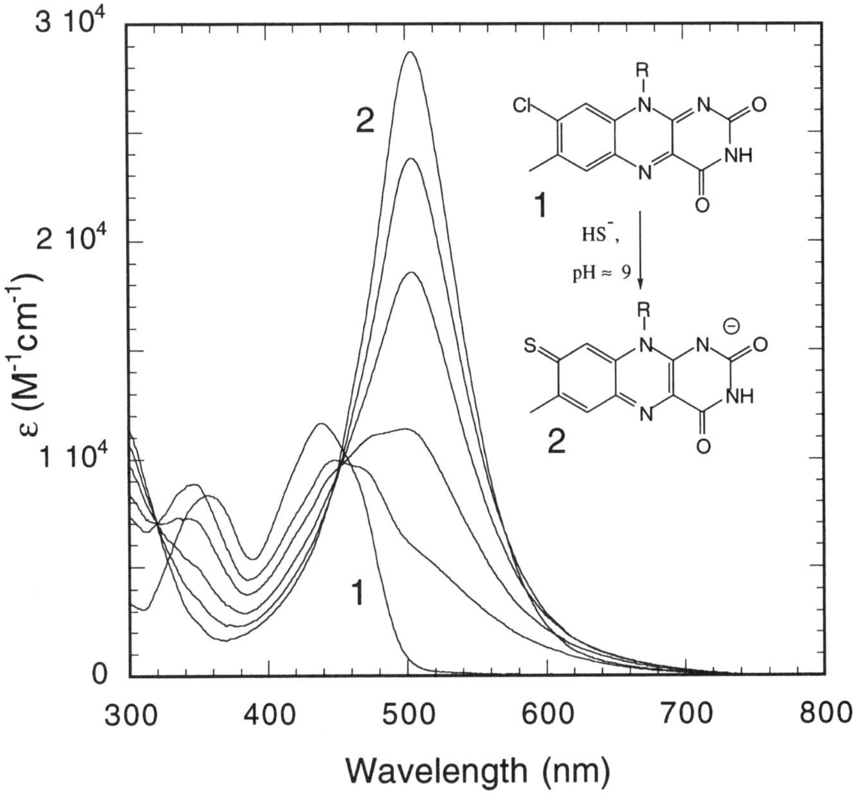

Fig. 4. Spectral changes accompanying the conversion of 8-Cl-flavin into 8-thioflavin (in its anionic form). The reaction was carried out anaerobically at 25°C, and in the presence of ≈10 m*M* of the reactant.

tron donation and uptake. Accessibility to these positions is to be found in those enzymes which carry out chemistry with a substrate (as opposed to the cases where the flavin exhibits a structural or electron transfer role). Accessibility to position C(4a) appears to be crucial for the formation of reactive oxygen hydroperoxides. Strategies to probe this specific point are difficult to conceive and have not yet been implemented.

The xylene ring is regarded as being "unreactive," mainly because there are no cases in which chemical events resulting from interaction with substrate have been found. Often this moiety is inserted into hydrophobic protein pockets. However, an unexpected feature has emerged: position C(8)-CH_3 appears to be near the protein surface in many flavoenzymes, in which it has no (detectable) function. In several cases it is freely accessible to bulk solvent, while other parts of the molecule are not. Massey's group has studied this aspect

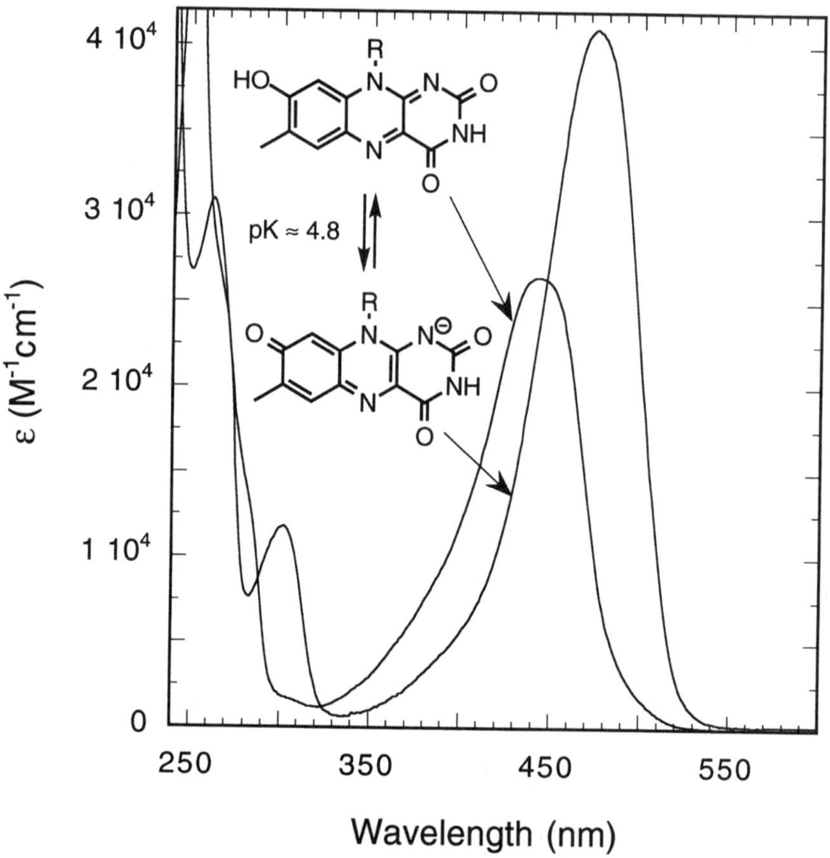

Fig. 5. Spectral changes accompanying ionisation of 8-OH-flavin. Such changes can be used to monitor e.g., the charge distribution at the active center on an enzyme, or p*K* shifts induced by the latter.

using the reactivity of 8-mercapto-flavins bound to various enzymes with alkylating electrophiles in the presence and absence of substrate ***(13)***. Accessibility to position C(8)-CH_3 can also be probed using 8-Cl-8-nor-flavins by following its reaction with sulfur nucleophiles. These approaches are easily implemented experimentally since they are based on the large spectral changes accompanying the conversions. Two examples are shown in **Figs. 4** and **5**, where the spectral effects accompanying the conversion of 8-Cl-8-nor-flavin into the (8)-S-anion, (**Fig. 4**), and that of the latter into its neutral form are shown (**Fig. 5**) Note that the absorption spectrum of the 8-SH and 8-S-alkyl flavins are very similar.

The flavin nucleus is also a sensitive tool for studying the environment surrounding it.

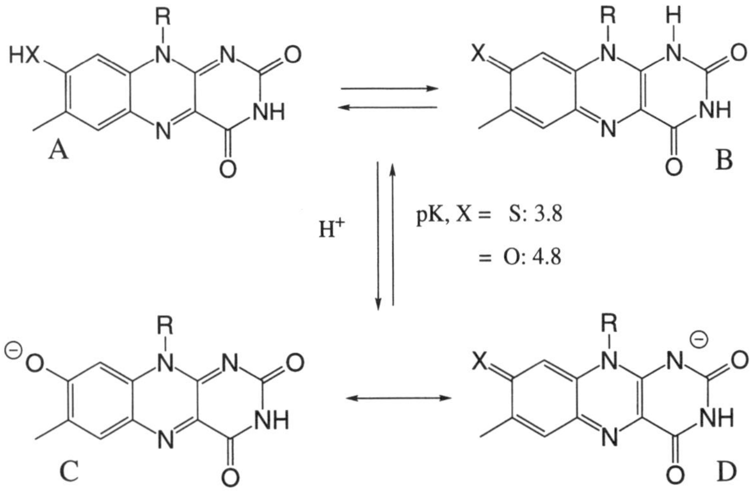

Fig. 6. Possible forms of 8-substituted flavins. Neutral species can exist either in the "phenolic" form (A), or in the "paraquinoid" form (B). These are spectrally distinct species. Ionisation leads to a delocalized system in which the charge can be delocalized in the pyrimidine moiety as shown by the canonical structure (D), the one which predominates in free solution. Stabilization of the "phenolate" structure (C) can be induced by binding to a protein active site.

The shape of the absorption spectrum of normal flavin is somewhat dependent on the solvent, the differences probably reflecting the relative intensities of single vibronic transitions, and being best detected by difference spectroscopy. Such a study has been presented in 1973 by Massey's group ***(14)*** in which also the temperature dependence of the flavin spectrum has been reported.

Binding of the flavin chromophore can be accompanied by very large perturbations of the spectrum as, e.g., in the case of cholesterol oxidase, where two members of the same family exhibit very different absorbance spectra ***(15)***. The direct interpretation of such effects is, however, very difficult, if not impossible in the case of flavoproteins. More information about the protein environment can be obtained using flavins in which the charge distribution is very different. This is the case with 8-substituted-8-nor-flavins, where the substituent is -NR_2, -SH, or OH. These molecules can exist in two canonic states as represented in **Fig. 6**, and where the charge (or the X-H bond) is either at position N(1) or on C(8)-X. Obviously these species are spectrally very different. With neutral 8-SH and 8-OH, the predominant form is a "phenolic" one, as shown on the left hand side on **Fig. 6**. The spectra are accordingly similar to those of normal flavins with maxima in the 450 nm region (**Fig. 7**). With 8-NH_2-8-nor-flavins, and with the anionic forms of 8-S-, and 8-O-8-nor-flavins the paraquinoid structure shown in **Fig. 6** (right hand side structure)

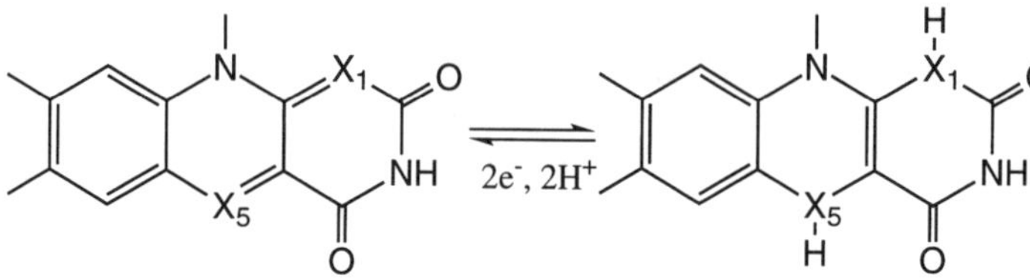

Fig. 7. Positions at which an isoelectronic exchange of "N" for "C-H" has been implemented in the flavin system.

predominates. This has been shown based on model work in which the respective forms can be blocked by alkylation ***(16,17)***. It should be pointed out, that in the anionic forms just named, the negative charge is located predominantly at the N(1)-C(2)=O locus. These flavins are thus, from the point of view of charge distributions, analogues of *reduced* flavins, where the negative charge also sits at the same location. The stabilization of the reduced flavin negative charge is an important factor in determining the redox potential of the couple, in that a stabilization, corresponding to a pK decrease, is equivalent to an increase of redox potential. Such flavins have thus been employed in the detection of a negative charge near N(1) e.g., in the L-lactate oxidase, the glycolate oxidase family ***(18)***, or with glutathione reductase, where an induced dipole brings about the same effect ***(19)***. The opposite, i.e., the destabilization of the paraquinoid form of anionic 8-S- or 8-O-flavins in favor of the phenolic mesomeric forms occurs with flavodoxins ***(17)***, which is, in turn, nicely correlated with the low redox potential of these proteins. **Figure 7** depicts an example of the effect of the 8-OH ionization on the chromophoric system of 8-hydroxy-8-nor-flavin.

3.2. Flavin Analogues as Mechanistic Probes

The outstanding property of the flavin ring is its chemical versatility which leads to a capacity to carry out differing reactions by intrinsically different mechanisms. This, in turn, can make the determination of a specific mechanistic variant very difficult. For answering this question flavin analogues can be tailored to carry out only specific reactions, while not allowing others. The modifications which have been most fruitful in this respect are based on the isoelectronic substitution of N with a C-H, which has been implemented at the flavin positions N(1) and/or N(5). The structures of these analogues are shown in their corresponding oxidized and 2 e- reduced forms (**Fig. 8**).

Four general questions can be addressed using these models; based on the following diversities:

1. Formation of acceptor H-bridges (*see* also **Fig. 2**). This is possible with *N*-H but not with C-H.

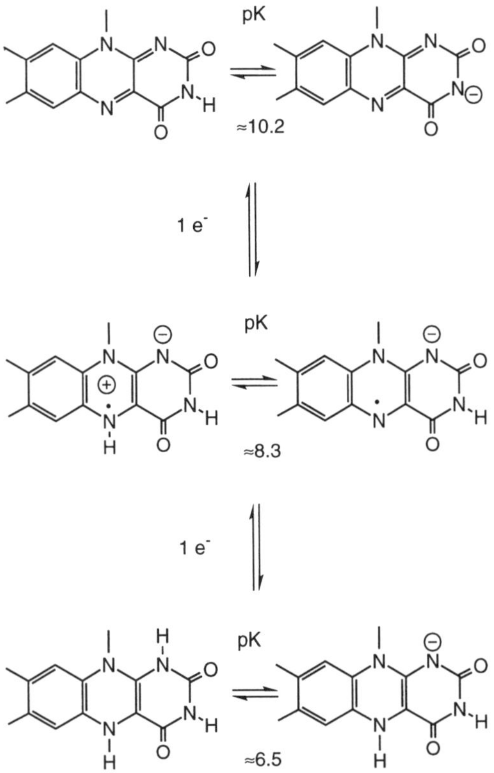

Fig. 8. Relevant structures of oxidized, "half-reduced" (semiquinone) and "fully reduced" (dihydro) flavins, and corresponding ionisations of biological relevance.

2. Formation of donor H-bridges: Possible with the N lone pair, not with C-H.
3. Occurrence of formation/rupture of a *N*-H bond during catalysis, and stereospecificity. The *N*-H bond is intrinsically labile, the C-H bond is stable. This difference can lead to mechanistic differentiations (e.g., in a transition state and in analysis of product stereospecificity).
4. Mode of rupture of a *N*-H bond during catalysis: Radical states are formed with much lower energy on N (and thus are more stable) than on a C-H center. Consequently, 5-deaza-5-carbaflavins have proven to be inactive in 1e- electron transfer processes.
5. Occurrence of covalent intermediates in catalysis. Such species would be expected to be stable when involving a C-C bond, as opposed to labile (i.e., catalytically viable) for the *N*-C case (one C originating from the flavin, the second from a reactant).

These properties have led to the study of 5-deaza-5-carba-FMN or -FAD as replacements of the normal flavin coenzymes in flavoenzymes and have proven to be mechanistically valuable flavin analogues. As stated above, this analogue is an obligatory two-electron acceptor/donor due to the unfavorable energetics involved in (transient) semiquinone formation. A survey of the literature shows the flavin to be functional in H^- (hydride) transfer reactions. Based on the reactivity with pyridine nucleotide reductants, this analogue has been suggested (probably in a strong oversimplification) to be a "flavin-shaped pyridine nucleotide analogue" ***(20)***. No activity is observed in systems expected to occur by one-electron transfer steps and no evidence for any covalent intermediates in flavoenzyme catalysis have been observed. The reduced flavin analogue is unreactive with O_2 which is expected mechanistically, as the reaction with normal flavin hydroquinone involves a one-electron transfer to form $O_2{\cdot}^-$-flavin semiquinone radical pair as the initial reaction step ***(21)***.

The 5-deaza-5-carbaflavin analogue, at the FMN level, was first introduced by Edmondson et al. ***(22)*** for studies with flavodoxin. These initial studies demonstrated that it binds to the protein but that it does not form a stable semiquinone as found with other FMN analogues with N at the 5-position. The analogue is capable of accepting H^- from pyridine nucleotides in several nicotinamide-adenine-dinucleotide phosphate (NAD(P)H) dependent flavoenzyme reductases or from H^- donors such as BH_4^-. Hydride donation to 5-deaza-5-carbaFAD has been demonstrated with medium chain acyl-CoA dehydrogenase, which catalyzes the αβ-dehydrogenation of acyl CoA thioesters ***(23)***. Also of basic importance are the observations of incorporation of substrate-derived H-label into the 5 position of 5-deaza-5-carbaflavin coenzymes (suggesting a H^- mechanism) with the following enzymes: D-amino acid oxidase ***(24,25)***, lactate oxidase ***(26)***, and flavocytochrome b_2: lactate dehydrogenase ***(27)***. All these enzymes have long been thought to function by a H^+ abstraction from the α-C-H of the substrate as the mechanism of C-H bond cleavage. These observations would be in keeping with recent structure-activity studies of D-amino acid oxidase ***(28)*** and L-lactate oxidase ***(29)*** which also are compatible with a H^- mechanism. Furthermore the 3-dimensional crystal structure of D-amino acid oxidase ***(30)*** and the properties of mutants of the same enzyme in which a potential H^+-abstracting base is absent, provide additional evidence in support of a H^--transfer mechanism. An alternative possibility that requires consideration (and further study) is the suggestion that substitution of 5-deaza-5-carbaflavin for the normal flavin enforces a change in mechanism and thus a different mode of C-H bond cleavage ***(26,31)***.

The 5-deazaflavin analogue 8-hydroxy-8-nor-5-deaza-5-carbaflavin has been shown ***(32)*** to be a valuable probe to determine the stereochemistry of H^- transfer from pyridine nucleotide to the enzyme-bound flavin. Five different pyridine nucleotide-dependent enzymes (glutathione reductase, mercuric reductase, *p*-hydroxybenzoate reductase, melilotate hydroxylase, and anthranilate hydroxylase) have all been shown to be reduced at the *re*-face of the bound flavin (by H^- transfer) in agreement with predictions from crystal structures of two of the enzymes tested *(31)*. In the case of medium chain acyl-CoA dehydrogenase transfer of the substrate β-H also occurs onto the flavin *re*-face ***(32)***.

5-Deaza-5-carbaflavins have been particularly useful in the assessment of the oxygen reactivity of reduced species ***(33)***. 1-Carba-1-deaza-FMN has been used successfully to refute the proposal ***(34)***, that in bacterial luciferase the emitter is an N(1) protonated flavin cation. This conclusion was based on the finding that the 1-carba-1-deaza-FMN luciferase complex is competent in oxygen activation and light emission, but cannot protonate at position 1 ***(34)***.

Another mechanistic approach where 8-substituted flavin analogues have been profitably used is in the application of linear free energy relationships for testing proposed mechanisms. Two types of enzymatic mechanism have been successfully tested. With bacterial luciferase the mode of generation of the excited state (light-emitting species) has been shown to proceed via a process in which charge is donated from the reduced flavin to the reactant subspecies, probably the aldehyde hydroperoxihemiacetal ***(35)***. A series of 8-substituted FMNs was used with bacterial luciferase and the rate of light emission decay was correlated with the 1e^- oxidation potentials of a corresponding series of 8-substituted lumiflavin models. The results were concluded to be compatible with a CIEEL type mechanism. With *p*-hydroxybenzoate hydroxylase the reactivity of flavin 4a-hydroperoxides in catalytic substrate monooxygenation was correlated with the p*K*a of the 4a-OH products perturbed by the inductive electronic effect from the 8-substituent using various 8-substituted flavins ***(36)***. This p*K*a was assumed to be linearly correlated with the Eox of the corresponding oxidized flavin. Similar 8-substituted flavin models have been used in Bruice's laboratory ***(37)*** to probe an electrophilic aromatic substitution mechanism in flavin-mediated hydroxylations. Indeed the flavin-4a-hydroperoxide was deduced to be the hydroxylating agent since the rate of hydroxylation is correlated to the inductive effect of the 8-substituent on the flavin which modulates the p*K* (and stability) of the resulting flavin-4a-hydroxy anion.

3.3. Oxidation-Reduction Potentials of Flavin Analogues

Because flavoenzymes catalyze either one- or two-electron transfer steps in catalysis, the ability to alter the energetics of these reactions by modification of the flavin ring is a valuable mechanistic probe. The protein environment of the bound flavin can modify the flavin redox potential in three ways: (1) through donating or accepting H-bonding to the pyrazine and pyrimidine rings and to N(5) (*see* **Fig. 2**), (2) by suitable placement of charge proximal to flavin ring positions that develop a charge on one- or two-electron reduction; the charge(s) can be positive or negative, thus stabilizing or destabilizing the (transient) species, and (3) by insertion of (parts) of the flavin ring(s) into

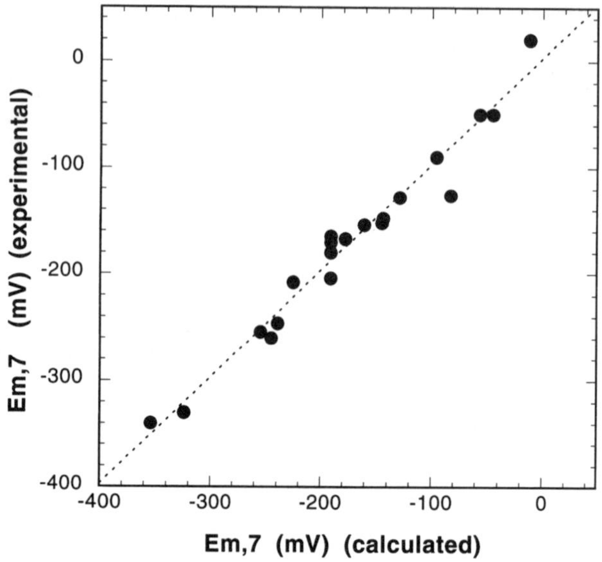

Fig. 9. Correlation of calculated and experimental redox potentials of the flavins listed in Table 2. The correlation has a slope of 1. See text for further details.

environments of varying dielectric (hydrophobic/hydrophilic) can have similar effects. The normal oxidized flavin ring is electrically neutral but develops charges on semiquinone formation to either the zwitterionic neutral (blue) or the anionic (red) semiquinone (pK = 8.4) forms (**Fig. 9**). The flavin hydroquinone form can either be uncharged or anionic (ionization at the N(1) position) with a pK of 6.5. Thus, anything that will affect the stability of charges (alteration of pK), and the [H^+] will affect the thermodynamics of electron transfer in the free system. The same effect can be obtained by chemical alteration of the flavin ring by the introduction of appropriate substituents and at various positions. A substantial shift of the redox potential occurs upon binding to the target enzyme.

To document the influence of flavin alteration on the two-electron redox potential, a list of 35 analogues and their respective $E_{m,7}$ values taken from the available literature are given in **Table 1**. These values range from a decrease (relative to FAD) of –30 mV to an increase of +230 mV or an overall range of 360 mV. The range for Ox/SQ and SQ/HQ potentials for those analogues listed in **Table 1** are not well documented, although the OX/SQ potential of the 5-deaza-5-carba-FAD has been determined to be extremely low (–700 mV *(38)*, which would remove this analogue from consideration of one-electron redox transfers due to the unfavorable energetics involved (*see* above). The alterations in potential can be predicted by the inductive influence of the substitu-

Table 1
Oxidation-Reduction Potentials of Flavin Analogues

Flavin analogue	Potential[a] ($E_{m,7}$) (mV)	Reference
1. FAD	–208	*(40)*
Pyrimidine ring-substituted analogues		
2. 4-ThioFAD	–55	*(41)*
3. 2-ThioFAD	–126	*(11)*
4. 3-Deaza-3-carba-FAD	–240	*(1)*
5. 1-Deaza-1-carba-FAD	–280	*(43)*
Pyrazine ring-substituted analogues		
6. 5-Deaza-5-carba-FAD	–310	*(1)*
7. 5-Deaza-5-thia-lumiflavin	+380 (SQ/HQ couple)	*(44)*
Benzenoid ring-substituted analogues		
8. 7,8-a,a,a-Hexafluororiboflavin	+20	*(45)*
9. 8-Fluoro-8-norFAD	–167	*(35)*
10. 8-Bromo-8-norFMN	–148	*(35)*
11. 7,8-Dichloro-7,8-norFAD	–126	*(46)*
12. 7-Chloro-7,8-norFAD	–128	*(46)*
13. 8-Chloro-8-norFAD	–152	*(47)*
14. 7-Bromo-7-norFAD	–154	*(48)*
15. 8-Mercapto-8-norFAD	–290	*(16)*
16. 8-Thiomethyl-8-norFAD	–204	*(35,49)*
17. 8-Methylthiosulfonyl-8-norFMN	–50	*(35)*
18. 8-Methylthiosufinyl-8-norFMN	–161	*(35)*
19. 8-Hydoxy-8-norFAD	–340	*(50)*
20. 8-Amino-8-norFAD	–330	*(35)*
21. 8-Dimethylamino-8-norFAD	–254	*(36)*
22. 8-Methoxy-8-norFMN	–260	*(35)*
23. 8-Ethoxy-8-norFMN	–246	*(35)*
24. 8-H-8-norFMN	–180	*(35)*
25. 6-Methyl-8-norFMN	–200	*(35)*
26. 6-Methylriboflavin	–219	*(51)*
27. 6-HydroxyFAD	–255	*(50)*
28. 6-ThiocyanatoFAD	Not known	
29. 6-AminoFAD	–297	*(50)*
30. 6-MercaptoFAD	Not known	
31. 9-Aza-riboflavin	–135	*(51)*
32. 8-Cyano-8-norriboflavin	–50	*(52)*
33. 8-Formyl-8-norriboflavin	–90[b]	*(53)*
34. 8-Carboxy-8-norriboflavin	–165	*(54)*
35. 8a-Hydroxyriboflavin	–170	*(54)*

(continued)

Table 1 *(continued)*

[a]All potentials listed are for the two-electron couple of the unbound flavin analogue vs the normal Hydrogen electrode unless otherwise stated.

[b]The listed potential is for the form in which the 8-formyl group is uncomplexed. Intramolecular hemiacetal formation with the riboflavin side chain hydroxyl group lowers the potential to –159 mV *(53)*.

ent on the lowest unoccupied molecular orbital (LUMO) of the oxidized flavin or on the highest occupied molecular orbital (HOMO) of the reduced flavin. A linear dependence of LUMO energy with oxidation-reduction potential has been well-documented with several other classes of redox-active compounds *(39)*. Alterations of these energies and the distribution of electron densities about the flavin ring can provide important insights into the reaction mechanism catalyzed.

Although modifications at just about all positions of the flavin ring will affect the redox potential, substitutions in the pyrimidine moiety will also drastically affect the chemistry of the flavin, and the interaction with the protein. Their use has thus been limited by these considerations. Because there are no documented specific interactions (H-bonding or charge interactions) of the benzene moiety of the flavin with the protein environment, its modification is less likely to alter or result in a protein-induced effect. Consequently, chemical substitutions at positions 6–8 have proven to cause less interference with flavin-protein interactions. In view of these considerations, redox-based studies of flavins modified at these positions (which are also chemically more accessible) have been favored. In order to convey an understanding of the role of electron donating-withdrawing groups in the 7 and 8 positions, which are considered to be respective *meta* and *para* positions to the N(5)-C(4a) reactive center of the flavin ring, we have carried out a correlation analysis of two-electron oxidation-reduction potential with the sum of Hammett σ values for substituents in these two positions. The oxidation-reduction potentials are taken from 20 of the analogues listed in **Table 1** and the values for σ (*para*) (8-substituent) and σ (*meta*) (7 position) are taken from *(55)*. Analogues with substituents in the pyrimidine ring and in the 6 and 9 positions of the benzene ring were not considered since the electronic effects of substituents, as in these positions are difficult to evaluate and also would exhibit steric contributions. An excellent correlation is observed which is described by the equation:

$$E_{m,7} = 182.7\ (\sigma_p+\sigma_m) - 185.7\ (R^2 = 0.94) \qquad (1)$$

This correlation does not include the 8-thio-8-nor-flavin, as the σ value for the anionic thiol substituent is not well characterized. We find that an average of the σ values for SH and S^- provides a reasonable fit which suggests the possibility that the p*K* of the 8 thiol group is altered from its value of 3.8 in the oxidized form to a higher value in the reduced form. No information in the literature is available on the p*K* of the 8-SH in the reduced flavin analogue. The 8-CH_3SO-analogue was also omitted since the potential may be somewhat altered by the slow reduction of the sulfoxide to the thioether during the redox potential measurement. Similar approaches have been suggested in ***(56)*** and in the course of writing this manuscript, we became aware of a similar study of the effect of 7 and 8 substitutions on the redox properties of lumiflavin analogues as measured by cyclic voltammetry ***(57)***. The results given in ***(57)*** are similar to the parameters given in **Eq. 1**.

The coefficient (or ρ value) in Eq. 1 is the sum of ρ_p and ρ_m and assumes they contribute equally, which is not necessarily true for a molecular system with the complexity of the isoalloxazine ring. The relative contributions of the sigma values for the 7 and 8 position substituents to the flavin redox potential can be separately evaluated by a two-component multiple regression statistical analysis in accord with the following equation and using the values listed in **Table 2**:

$$E_{m,7}\ (X_p, Y_m) = \rho_p\ \sigma_p + \rho_m\ \sigma_m + E_{m,7}(H,H) \qquad (2)$$

The results for this analysis are given below:

$$\rho_p = 203 \pm 10\ \mathrm{mV}/\ \sigma_p$$
$$\rho_m = 133 \pm 21\ \mathrm{mV}/\sigma_m$$
$$E_{m,7}(H,H) = -178 \pm 4\ \mathrm{mV}$$

The statistical values for this correlation are: $F_{2,19} = 242$ ($P<0.001$), $R^2 = 0.97$

The analysis provides a more valid determination of ρ_p than of ρ_m, as there are more examples of 8- than of 7-substitutions for the flavin analogues listed in **Table 2**. The results given above also demonstrate that the electronic contribution of the 7 substituent is one-third less than that of the 8 substituent. This analysis is of value for its ability to predict which substituents in the 7 and/or 8 positions of the flavin ring would provide a flavin analogue with the desired potential. To illustrate this point, the calculated redox potentials of the 20 flavin analogues listed in **Table 2** were estimated using **Eq. 2** and the parameters listed below it. On substitution of the appropriate σ values for the *para* and *meta* substituents, the redox potentials are readily calculated. Comparison of the calculated and experimental potentials are given in **Table 2** and shown in a graphical form in **Fig. 9**. A linear relationship exists with an expected slope of 1, an intercept (at 0 potential) of zero, and an R^2 value of 0.99. Thus, using **Eq. 2** and the given parameters, it is possible to estimate the potential of any 7 and/or 8 substituted flavin analogue with a high degree of accuracy.

In practice, however, the influence of the protein environment considerably modulates the one- and two-electron potentials of the bound flavin and may

Table 2
Compilation of Sigma Values and Experimental and Calculated Potentials of 7,8-Substituted-Flavin Analogues

Flavin Substituent[a]		Sigma Values		Potentials		
				(mV)[b]	(mV)[c]	
8	7	*para*	*meta*	(expl)	(calc)	No.[d]
CH_3	CH_3	–0.170	–0.069	–208	–225	1
CF_3	CF_3	0.54	0.43	+20	–12	8
F	CH_3	0.062	–0.069	–167	–178	9
Br	CH_3	0.232	–0.069	–148	–144	10
Cl	Cl	0.227	0.373	–126	–83	11
H	Cl	0	0.373	–128	–129	12
Cl	CH_3	0.227	–0.069	–152	–145	13
CH_3	Br	–0.17	0.391	–154	–161	14
S^-	CH_3	–1.21	–0.069	–290	—	15
CH_3S	CH_3	0	–0.069	–204	–191	16
CH_3SO_2	CH_3	0.720	–.069	–50	–45	17
CH_3SO	CH_3	0.49	–0.069	–161	—	18
O^-	CH_3	–0.81	–0.069	–340	–354	19
NH_2	CH_3	–0.660	–0.069	–330	–324	20
Me_2N	CH_3	–0.32	–0.069	–254	–255	21
MeO	CH_3	–0.268	–0.069	–260	–245	22
EtO	CH_3	–0.240	–0.069	–246	–239	23
H	CH_3	0	–0.069	–180	–191	24
CN	CH_3	0.660	–0.069	–50	–57	32
CHO	CH_3	0.47	–0.069	–90	–96	33
COO^-	CH_3	0	–0.069	–165	–191	34
CH_2OH	CH3	0	–0.068	–170	–191	35

[a]Sigma values were taken from tables in Hansch and Leo *(55)*.
[b]The experimental $E_{m,7}$ values for each flavin analogue are taken from **Table 1**.
[c]The calculated $E_{m,7}$ values for each flavin analogue are from **Eq. 2**.
[d]The given numbers for each substituted flavin analogue correspond to those in **Table 1**.

differentially alter those of the bound flavin analogue relative to that of the normal flavin due to influences on the expected differential p*K* values of the redox (SQ and/or HQ) forms of the analogue. This perturbation thus provides additional information on the environment of the flavin binding site. In spite of these additional considerations which makes direct correlations of enzyme rates and other catalytic parameters with the redox potential of the free flavin analogue a dubious exercise, there are examples in the literature where successful correlations have been made including: bacterial luciferase *(35)*, xanthine oxidase ***(58)***, and pyridine nucleotide-dependent flavin reductases ***(51)***. The number of potential systems that could be studied using flavin analogues with a range of potentials is considerable. A rigorous application to an enzyme system requires not only the influence of the flavin analogue on reaction rates but also requires determination of the potential of the bound analogue under experimental conditions as done by Massey and colleagues to correlate the spectral properties of the charge transfer complex of the FMN of Old Yellow Enzyme with *p*-hydroxybenzaldehyde with the redox potential of the bound flavin analogue ***(59)*** and to probe the reactivity of flavin-4a-peroxy intermediates in *p*-hydroxybenzoate hydroxylase ***(36)***.

References

1. Spencer, R., Fisher, J., and Walsh, C. (1976) Preparation, characterization, and chemical properties of the flavin coenzyme analogues 5-deazariboflavin, 5-deazariboflavin 5'-phosphate, and 5-deazariboflavin 5'-diphosphate, 5' leads to 5'-adenosine ester. *Biochemistry* **15,** 1043–1053.
2. Kim, J., Fuller, J. H., Kuusk, V., Cunane, L., Chen, Z., Mathews, F. S., and McIntire, W. S. (1995) The cytochrome subunit is necessary for covalent FAD attachment to the flavoprotein subunit of P-cresol methylhydroxylase. *J. Biol. Chem.* **270,** 31,202–31,209.
3. Miller, J. R. and Edmondson, D. E. (1997) Effect of flavin structure on the enzymatic activity of recombinant human liver monoamine oxidase A, in *Flavins and Flavoproteins* (Stevenson, K., Massey, V., and Williams, Jr., C. H., eds.), University of Calgary Press, Calgary, Alberta, Canada, pp. 71–75.
4. Engst, S., Vock, P., Wang, M., Kim, J. J., and Ghisla, S. (1998). On the mechanism of activation of Acyl-CoA substrates by medium chain Acyl-CoA dehydrogenase: interaction of the thioester carbonyl with the FAD ribityl side-chain. *Biochemistry,* in press.
5. Muller, F. and Hemmerich, P. (1966) Thiones, imines, oximes and azines of riboflavin. Nucleophilic substitution reactions in the flavin nucleus. Studies in the flavin series. X. *Helv. Chim. Acta* **49,** 2352–2364.
6. Lambooy, J. P. (1971) Analogs of riboflavin. *Meth. Enzymol.* **18,** 437–447.
7. Lambooy, J. P. (1967) The alloxazines and isoalloxazines, in *Heterocyclic Compounds* (Elderfield, R. C., ed.), vol. 9, 2, Wiley, New York.
8. Hausinger, R. P., Honek, J. K., and Walsh, C. (1986) *Enzymol.* **122,** 199–220.

9a. Tsibris, J. C. M., McCormick, D. B., and Wright, L. D. (1966) Studies on the binding and function of flavin phosphates with flavin mononucleotide-dependent enzymes *J. Biol. Chem.* **241,** 1138–1143.

9b. Chassy, B. M. and McCormick, D. B. (1965) Coenzyme specificity of D-amino acid oxidase from the flavin moiety of FAD. *Biochim. Biophys. Acta* **110,** 91–96.

10. Muller, F. and Hemmerich, P. (1966) Thione, imine, oxime und azine des riboflavins. Nucleophile substitutionsreaktionen am flavinkern. *Helv. Chim. Acta* **49,** 2352–2364.

11. Claiborne, A., Hemmerich, P., Massey, V., and Lawton, R. (1983) Reaction of 2-thio-FAD-reconstituted p-hydroxybenzoate hydroxylase with hydrogen peroxide. Formation of a covalent flavin-protein linkage. *J. Biol. Chem.* **258,** 5433–5439.
12. Biemann, M., Claiborne, A., Ghisla, S., and Massey, V. (1984) 4-Thioflavins as active site probes of flavoproteins. Reactions with sulfite. *J. Biol. Chem.* **259,** 13,355–13,362.
13. Schopfer, L. M., Massey, V., and Claiborne, (1981) Active site probes of flavoproteins. Determination of the solvent accessibility of the flavin position 8 for a series of flavoproteins. *J. Biol. Chem.* **256,** 7329–7337.
14. Muller, F., Mayhew, S. G., and Massey, V. (1973) On the effect of temperature on the absorption spectra of free and protein-bound flavins. *Biochemistry* **12,** 4654–4662.
15. Gadda, G., Wels, G., Ambrosius, D., Pilone, M. S., and Ghisla, S. (1997) Characterization of cholesterol oxidase from *Streptomyces hygroscopicus* and *Brevibacterium sterolicum. Eur. J. Biochem.* **250,** 360–376.
16. Massey, V., Ghisla, S., and Moore, D. G. (1979) 8-Mercapto-flavins as active site probes of flavin enzymes. *J. Biol. Chem.* **254,** 9640–9650.
17. Ghisla, S. and Mayhew, S. G. (1976) Identification and properties of 8-hydroxy-flavin adenine dinucleotide in electron-transferring flavoprotein from *Peptostreptococcus elsdenii. Eur. J. Biochem.* **63,** 373–390.
18. Ghisla, S. and Massey, V. (1991) L-Lactate oxidase, in *Chemistry and Biochemistry of Flavoenzymes* (Muller, F., ed.), vol. II, CRC Press, Boca Raton, FL, pp. 243–289.
19. Schultz, G. E., Schirmer, F. H., and Pai, E. F. (1982) FAD-binding site of glutathione reductase. *J. Mol. Biol.* **160,** 287–308.
20. Hemmerich, P., Massey, V., Michel, H., and Schug, C. (1982) Scope and limitation of single electron transfer in biology. *Structure Bonding* **48,** 93–123.
21. Eberlein, G. and Bruice, T. C. (1982) One- and two-electron reduction of oxygen by 1,5-dihydroflavins. *J. Am. Chem. Soc.* **104,** 1449–1452.
22. Edmondson, D. E., Barman, B., and Tollin, G. (1972) On the importance of the *N*-5 position in flavin coenzymes. Properties of free and protein-bound 5-deaza analogs. *Biochemistry* **11,** 1133–1137.
23. Ghisla, S., Thorpe, C., and Massey, V. (1984) Mechanistic studies with general acyl-CoA dehydrogenase and butyryl-CoA-dehydrogenase, evidence for the transfer of the β-hydrogen as a Hydride. *Biochemistry* **23,** 3154–3161.
24. Hersh, L. B. and Jorns, M. S. (1975) Use of 5–deazaFAD to study hydrogen transfer in the D-amino acid oxidase reaction. *J. Biol. Chem.* **250,** 8728–8734.
25. Fisher, J., Spencer, R., and Walsh, C. (1976) Enzyme-catalyzed redox reactions with the flavin analogues 5-deazariboflavin, 5-deazariboflavin 5'-phosphate, and 5-deazariboflavin 5'-diphosphate, 5' leads to 5'-adenosine ester. *Biochemistry* **15,** 1054–1064.
26. Averill, B. A., Schonbrunn, A., Abeles, R. H., Weinstock, L. T., Cheng, C. C., Fisher, J., Spencer, R., and Walsh, C. (1975) Studies on the mechanism of *Mycobacterium smegmatis* L-lactate oxidase. 5-Deazaflavin mononucleotide as a coenzyme analogue. *J. Biol. Chem.* **250,** 1603–1605.

27. Balme, A. and Lederer, F. (1993) Reconstitution of flavin-free flavocytochrome b2 with 5-deazaFMN: a carbanion or a hydride mechanism?, in *Flavins and Flavoproteins* (Yagi, K., ed.), W. DeGruyter & Co., Berlin, pp. 629–637.
28. Pollegioni, L., Blodig, W., and Ghisla, S. (1997) On the mechanism of D-amino acid oxidase: structure linear free energy correlations and deuterium kinetic isotope effects using substituted phenylglycines. *J. Biol. Chem.* **272,** 4924–4934.
29. Yorita, K., Sanko, K., Aki, K., Ghisla, S., Palfrey, B. A., and Massey, V. (1997) On the reaction mechanism of L-lactate oxidase: quantitative structure-activity analysis of the reaction with para-substituted L-mandelates. *Proc. Natl. Acad. Sci. USA* **94,** 9590–9595.
30. Mattevi, A., Vanoni, M. A., Todone, F., Teplyakov, A., Coda, A., Bolognesi, M., and Curti, B. (1997) Crystal structure of D-amino acid oxidase: a case of active site mirror-image convergent evolution with flavocytochrome, *Proc. Natl. Acad. Sci. USA* **93,** 7496–7501.
31. Ghisla, S. and Massey, V. (1986) New flavins for old: artificial flavins as active site probes of flavoproteins. *Biochem. J.* **239,** 1–12.
32. Manstein, D. J., Pai, E. F., Schopfer, L. M., and Massey, V. (1986) Absolute stereochemistry of flavins in enzyme-catalyzed reactions. *Biochemistry* **25,** 6807–6816.
33. Tittmann, K., Proske, D., Spinka, M., Ghisla, G., Rudolph, R. G. H., and Kern, G. (1998) Activation of thiamin diphosphate and FAD in the phosphate dependent pyruvate oxidase from *lactobacillus plantarum. J. Biol. Chem.* **273,** 12,929–12,934.
34. Kurfurst, M., Macheroux, P., Ghisla, S., and Hastings, J. W. (1989) Bioluminescence emission of bacterial luciferase with 1-deaza-FMN: evidence for the noninvolvement of N(1)-protonated flavin species as emitters. *Eur. J. Biochem.* **181,** 453–457.
35. Eckstein, J. W., Hastings, J. W., and Ghisla., S. (1993) Mechanism of bacterial bioluminescence: 4a,5-dihydroflavin analogs as models for luciferase hydroperoxide intermediates and the effect of substituents at the 8 position of flavin on luciferase kinetics. *Biochemistry* **32,** 404–411.
36. Ortiz-Meldonado, M., Ballou, D. P., and Massey, V. (1997) Leaving group tendencies of 8-substituted flavin-C4a-alkoxides and the mechanism of hydroxylation catalyzed by p-hydroxybenzoate hydroxylase, in *Flavins and Flavoproteins* (Stevenson, K., Massey, V., and Williams, Jr., C. H., eds.) University of Calgary Press, Calgary, Alberta, Canada, pp. 323–326.
37. Bruice, T. C., Noar, J. B., Ball, S. S., and Venkataram, U. V. (1983) The monooxygen donation potential of 4a-hydroperoxy flavins as compared with percarboxylic acid and other hydroperoxides. Monooxygen donation to olefin, tertiary amine, alkyl sulfide, and iodide ion. *J. Am. Chem. Soc.* **105,** 2452–2462.
38. Blankenhorn, G. (1976) Nicotinamide-dependent one-electron and two-electron (flavin) oxidoreduction: thermodynamics, kinetics, and mechanism. *Eur. J. Biochem.* **67,** 67–80.
39. Streitwieser, Jr., A. (1966) *Molecular Orbital Theory for Organic Chemists,* Wiley, New York.

40. Draper, R. D. and Ingraham, L. L. (1968) A potentiometric study of the flavin semiquinone equilibrium. *Arch. Biochem. Biophys.* **125,** 802–808.
41. Massey, V., Claiborne, A., Biemann, M., and Ghisla, S. (1984) 4–Thioflavins as active site probes of flavoproteins: general properties. *J. Biol. Chem.* **259,** 9667–9678.
42. Claiborne, A., Massey, V., Fitzpatrick, P. F., and Schopfer, L. M. (1982) 2-Thioflavins as active site probes of flavoproteins. *J. Biol. Chem.* **257,** 174–182.
43. Spencer, R., Fisher, J., and Walsh, C. (1977) Reconstitution of flavin enzymes with 1-carba-1-deazaflavin coenzyme analogues. *Biochemistry* **16,** 3594–3602.
44. Fenner, H., Grauert, R., Hemmerich, P., Michel, H., and Massey, V. (1979) 5-Thia-5-deazaflavin, a 1e-transferring flavin analog. *Eur. J. Biochem.* **95,** 183–191.
45. Nielsen, P., Bacher, A., Darling, D., and Cushman, M. (1983) Synthesis and biological evaluation of 7α,α,α,8α,α,α-hexafluororiboflavin and 7α,α,α,8α,α,α-hexafluoro-FMN. *Z. Naturforsch.* **38c,** 701–707.
46. Massey, V. and Nishino, T. (1980) D-Amino acid oxidase containing 7,8-dichloro-FAD instead of FAD, in *Flavins and Flavoproteins* (Yagi, K., and Yamano, T., eds.), University Park Press, Baltimore, MD, pp. 1–11.
47. Moore, E., Ghisla, S., and Massey, V. (1979) Properties of flavins where the 8-methyl group is replaced by mercapto residues. *J. Biol. Chem.* **254,** 8173–8178.
48. Thorpe, C. and Massey, V. (1980) Flavin analogue studies of pig kidney general acyl-CoA dehydrogenase. *Biochemistry* **22,** 2972–2978.
49. Light, D. R. and Walsh, C. (1980) Flavin analogs as mechanistic probes of adrenodoxin reductase-dependent electron transfer to the cholesterol side chain cleavage cytochrome P-450 of the adrenal cortex. *J. Biol. Chem.* **255,** 4264–4277.
50. Massey, V. (1981) A simple method for the determination of redox potentials, in *Flavins and Flavoproteins* (Curti, B., Ronchi, S., and Zanetti, G., eds.), W. DeGruyter & Co., Berlin, pp. 59–66.
51. Walsh, C., Fisher, J., Spencer, R., Graham, D. W., Ashton, W. T., Brown, J., Brown, R. D., and Rogers, E. F. (1978) Chemical and enzymatic properties of riboflavin analogues. *Biochemistry* **17,** 1942–1951.
52. Murthy, Y. U. and Massey, V. (1998) Synthesis and properties of 8-CN-flavin nucleotide analogs and studies with flavoproteins. *J. Biol. Chem.* **272,** 8975–8982.
53. Edmondson, D. E. (1974) Intramolecular hemiacetal formation in 8-formylriboflavine. *Biochemistry* **13,** 2817–2821.
54. Edmondson, D. E. and Singer, T. P. (1973) Oxidation-reduction properties of the 8 α-substituted flavins., *J. Biol. Chem.* **248,** 8144–8149.
55. Hansch, C. and Leo, H. (1979) *Substituent Constants for Correlation Analysis in Chemistry and Biology,* Wiley, New York.
56. Macheroux, P., Eckstein, J., and Ghisla, S. (1987) Studies on the mechanism of bacterial bioluminescence. Evidence compatible with a one electron transfer process and a CIEEL mechanism in the luciferase reaction, in *Flavins and Flavoproteins* (McCormick, D. B. and Edmondson, D. E., eds.), DeGruyter, Berlin, pp. 613–619.

57. Hasford, J. J. and Rizzo, C. J. (1998) Linear free energy substituent effect on flavin redox chemistry. *J. Am. Chem. Soc.* **120,** in press.
58. Hille, R., Fee, J. A., and Massey, V. (1981) Equilibrium properties of xanthine oxidase containing FAD analogs of varying oxidation-reduction potential. *J. Biol. Chem.* **256,** 8933–8940.
59. Stewart, R. C. and Massey, V. (1985) Potentiometric studies of native and flavin-substituted Old Yellow Enzyme. *J. Biol. Chem.* **260,** 13,639–13,647.

13

Identification of Covalent Flavoproteins and Analysis of the Covalent Link

Nigel S. Scrutton

1. Introduction

Flavin is a versatile cofactor involved in a wide spectrum of chemical transformations in biology. It is therefore not surprising that the flavoenzyme family represents one of the largest collections of redox enzyme molecules. The so-called covalent flavoproteins form a subgroup of this larger family of flavoproteins *(1)* and, although a relatively small group (approx 25 members as of 1998), covalent flavoproteins have received considerable attention, both from the viewpoint of enzyme mechanism and also the mode by which the flavin redox center becomes attached to the protein scaffold. The covalent flavoproteins fall into two categories—those in which the flavin is attached at the C6 position of the flavin isoalloxazine ring and those where linkage is via the 8α methyl group (**Fig. 1**). The former group has only two members (tri- and dimethylamine dehydrogenases) and attachment occurs via a cysteine residue forming a C6-thioether linkage. The 8α methyl grouping forms the majority of the covalent flavoprotein family and members are typified by linkages between the 8α methyl group of the flavin isoalloxazine ring and histidine, tyrosine or cysteine side chains. The role of covalently bound flavins in redox enzymology has been the focus of much debate. For example, modulation of flavin reduction potential *(2,3)* and improvement of electron transfer rates to downstream redox acceptors by enhancing electronic coupling between cofactors *(4)* have been proposed. Recent work on trimethylamine dehydrogenase suggests that covalent linkage may have evolved to suppress hydroxylation of the isoalloxazine ring at the C6 position, thereby preventing inactivation of the redox center *(5)*. Analysis of a variety of covalent flavoproteins has revealed that formation of the covalent link is a self-catalytic process *(4,6,7)*, and mecha-

From: Methods in Molecular Biology, vol. 131: Flavoprotein Protocols
Edited by: S. K. Chapman and G. A. Reid © Humana Press Inc., Totowa, NJ

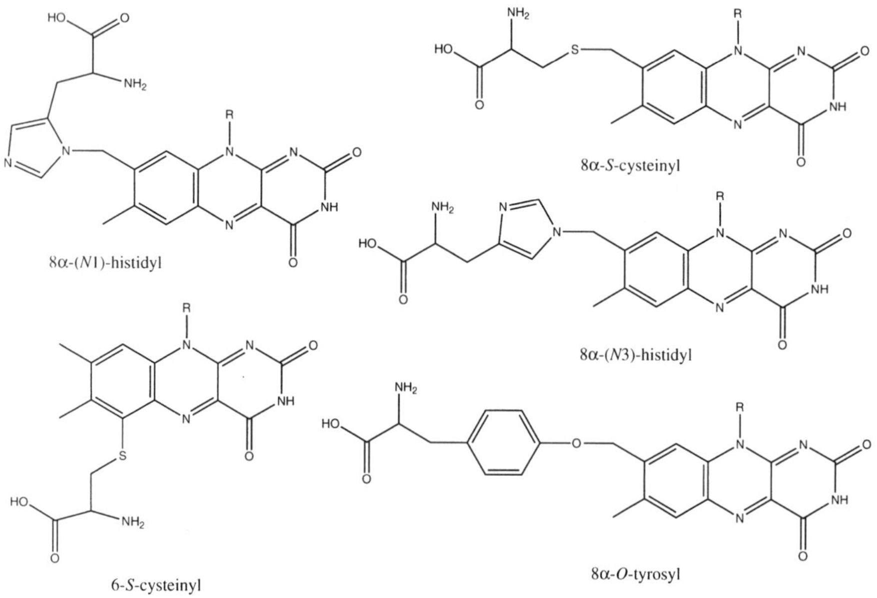

Fig. 1. Structures of the known flavin amino acids found in covalent flavoenzymes.

nisms for flavinylation have been proposed. For selected enzymes, studies by directed mutagenesis have confirmed the roles of key residues in the flavinylation reaction [e.g., ***(4,8)***].

This chapter describes standard methods used for identifying covalent flavoproteins and determining the precise nature of the chemical link. The methods can be carried out without substantial investment in state-of-the-art equipment, although more rapid identification of the link might be achieved where there is ready access to facilities such as mass spectrometry. Given the "low-tech" approach adopted in this article, the analytical methods are well suited to molecular biology and enzymology laboratories where there is no access to the "high-tech" environment of a modern protein chemistry laboratory.

2. Materials

As is the case generally for analytical techniques, all reagents used should be of the highest purity and all glassware should be carefully washed and rinsed. Water should be ultrapure [e.g., as from a Milli Q unit (Millipore)].

1. *Naja naja atra* snake venom (a source of phosphodiesterase), calf intestine alkaline phosphatase, porcine kidney aminopeptidase *M* (Sigma Chemical Co., St Louis, MO).
2. Precoated glass cellulose (microcrystalline avicel) thin-layer chromatography plates (Camlab Ltd., Cambridge, UK).

3. Pepsin, chymotrypsin, and trypsin (Boehringer Mannheim).
4. 5:1:5 (v/v/v) Methanol, acetic acid, water. 5:1:5 (v/v/v) methanol, acetic acid, water plus 0.2% Coomassie Brilliant Blue R250.
5. 7% Acetic acid.
6. 5% Formic acid.
7. Performic acid (obtained by incubating 9 vol formic acid with 1 vol 30% H_2O_2 for 1 h at room temperature).
8. 55% (w/v) Trichloroacetic acid.
9. 1% (w/v) Trichloroacetic acid.
10. Diethyl ether.
11. Acetone.
12. Phosphocellulose (fine mesh; Sigma).
13. 10 m*M* Sodium acetate, pH 6.0.
14. 20 m*M* Tris-HCl, pH 8.0, 1 m*M* $MgCl_2$.
15. 1 *M* Pyridine acetate, pH 5.0.
16. 100 m*M* Citrate-phosphate buffer, pH range spanning 2.5–8 in increments of 0.5 pH units.
17. Sodium dithionite and sodium borohydride (Sigma).

3. Methods

3.1. Identification of Covalent Flavoproteins

The identification of covalent flavoproteins is a relatively simple task and can be achieved by sodium dodecyl sulfate (SDS) polyacrylamide gel electrophoresis and subsequent detection of the flavin (by virtue of its yellow appearance or by fluorescence) associated with the protein. This simple procedure has been reported for many flavoprotein enzymes ***(9–11)***.

1. Electrophorese candidate covalent flavoprotein by conventional SDS-polyacrylamide gel electrophoresis. Apply three samples of the protein in neighboring wells of the polyacrylamide gel.
2. Following electrophoresis, with a clean razor blade dissect the gel so that each sample of the electrophoresed protein is contained in a separate gel portion.
3. Stain one of the gel pieces with Coomassie Brilliant Blue (R250) and subsequently destain for 30 min using the conventional procedure [5:1:5 (v/v) methanol, acetic acid, water].
4. Soak a second gel piece containing another sample of the protein with 7% acetic acid (about 15 min). The third gel piece is treated similarly, but after spraying with performic acid (*see* **Note 1**).
5. Illuminate the gel pieces prepared in **step 4** with ultraviolet light from a handheld transilluminator (λ 365 nm) to detect the intrinsic fluorescence of any covalently bound flavin. A comparison of the Coomassie-stained and UV-illuminated gel samples will identify which protein bands contain the covalently attached flavin.

3.2. Analysis of the Covalent Link

3.2.1. Strategy

Determination of the nature of the covalent link in flavin-containing peptides is not necessarily straightforward since physical-chemical properties such as fluorescence and the pK_a values of flavin titratable groups in flavinylated peptides will be influenced by the chemical composition of the peptide itself. Unequivocal identification of the chemical link involves proteolysis to produce a suitable flavinylated peptide, purification of the peptide and subsequent treatment with *Naja naja* snake venom (to convert the flavinylated peptides containing derivatives of FAD into the corresponding FMN form) and alkaline phosphatase (to dephosphorylate the FMN derivative to the riboflavin level). The aminoacyl flavin is then produced by treatment with aminopeptidase *M*, which is then purified further by e.g., ion-exchange chromatography, reverse-phase HPLC, or thin-layer chromatography (TLC). Having "released" the aminoacyl flavin from the peptide and thereby having avoided complications due to perturbation of the physico-chemical properties of the flavin by the peptide composition, the behavior of the aminoacyl flavin can then be compared with that of known flavin derivatives, thus enabling unequivocal identification of the linkage type. The general approach described here has been reported by Singer and McIntire ***(12)***. Since aminoacyl flavins (especially 6-*S*-cysteinyl flavins) are light sensitive, all procedures should be conducted in the dark.

3.2.2. Isolation of a Flavinylated Peptide

The first steps in the determination of the chemical link require the isolation and purification of a suitable flavinylated peptide. The precise method by which this is achieved will depend on the properties of the protein being investigated. For illustrative purposes, the method used for generating a flavinylated peptide from *Chlorobium* cytochrome c_{553} ***(13)*** is reproduced here, and similar approaches can be used for other covalent flavoproteins. Related methods using alternative proteases for isolating flavinylated peptides from other covalent flavoproteins have been reported [e.g., ***(9,10,14–18)***] and should be consulted prior to initiating a new investigation.

1. Pellet an ammonium sulfate suspension of *Chlorobium* cytochrome c_{553} (about 40 mg) by centrifugation (10 min at 33,000g, 4°C) and redissolve in water (1–2 mL).
2. Denature the protein by adding 0.1 vol of 55% (w/v) trichloroacetic acid at 0°C and collect the protein precipitate by centrifugation as in **step 1**. Wash the precipitate with 1% (w/v) trichloroacetic acid. Resuspend the protein in a minimal amount of water and remove residual trichloroacetic acid by extraction with 4 vol of diethyl ether. Remove residual ether by aspiration with nitrogen gas.

3. Add formic acid to a final concentration of 5% (v/v) and adjust the vol with 5% formic acid to a protein concentration of about 15 mg/mL.
4. Digest the protein with pepsin (performed in the dark) by adding 0.06 mg of pepsin per mg of the cytochrome (*see* **Note 2**). Digest for 4 h at 38°C.
5. Apply the digest directly to a column (0.9 × 12 cm) of phosphocellulose (pyridinium form) that has been equilibrated with 5% (v/v) formic acid. Elute column with 5% formic acid. Pool flavin containing fractions (identified by absorbance at 450 nm) and lyophilise to remove solvent (*see* **Note 3**).

At this point, the absorption spectrum of the peptide can be analyzed to aid identification, or alternatively analysis can await the production of the aminoacyl flavin described in **Subheading 3.2.3.**

3.2.3. Isolation of the Aminoacyl Flavin

Isolation of the aminoacyl flavin from the flavinylated peptide involves further proteolytic digestion with aminopeptidase *M* and dephosphorylation with *Naja naja* snake venom (a source of phosphodiesterase) and alkaline phosphatase. Again, variations in the procedure have been reported [e.g., ***(9–11, 13–21)***], but a typical method is given below. The extent of aminopeptidase *M* digestion can be assessed and purification of the aminoacyl flavin achieved using thin-layer chromatography. A good example of this approach has been reported by McIntire and coworkers ***(19)*** in their characterization of the covalent flavin of *p*-cresol methylhydroxylase as 8α-*O*-tyrosyl-FAD. Their purification method is also described here.

1. To convert an FAD peptide to the FMN level, incubate about 50 nmol of the flavinylated peptide with 200 μg/mL *Naja naja* snake venom in 10 m*M* sodium acetate, pH 6.0 for 2h at 25°C (*see* **Note 4**).
2. To convert the FMN peptide (produced in **step 1**) to the riboflavin level, add calf intestine alkaline phosphatase (140 units) and incubate for 2 h at 37°C in 20 m*M* Tris/HCl, pH 8.0, containing 1 m*M* $MgCl_2$.
3. Incubate the riboflavin peptide for 24 h at 38°C with 20 μg aminopeptidase *M* contained in 140 m*M* Tris/HCl pH 8.0. After 24 h, add an additional 20 μg aminopeptidase *M* and incubate for a further 12 h (*see* **Note 5**).
4. Lyophilize the aminoacyl riboflavin and resuspend in 25 μL of water (*see* **Note 6**).
5. Apply the aminoacyl flavin solution to an Avicel cellulose plate (10 × 20 cm) and perform chromatography in a 1-butanol/pyridine/acetic acid/water (15:3:10:12, v/v/v) solvent system. Scrape the major yellow fluorescent band (minor bands may also be seen if digestion is incomplete) from the plate, crush, suspend in acetone, wash in a column (1 cm diameter) with acetone and subsequently elute with 1.5 mL each of 1 *M* pyridine acetate, pH 5.0, and 5% (v/v) pyridine. Combine fractions (*see* **Note 7**).

Table 1
Spectroscopic Properties of the 6-*S*-Cysteinyl Flavins and the 8α Aminoacyl Flavins

Flavin derivative	Maximum wavelength of UV-visible absorption peaks (nm) and extinction coefficients (in parentheses; $mM^{-1}cm^{-1}$)				pK_a values for fluorescence changes as a function of pH
	λ_1	λ_2	λ_3	λ_4	
6-*S*-Cysteinyl riboflavin[a]	437	—	nd	nd	—
8α-*S*-Cysteinyl riboflavin[b]	448 (12.0)	367 (8.6)	270 (?)	nd	—
8α-*O*-Tyrosyl riboflavin[c]	446 (12.3)	359 (9.38)	268 (36.4)	224 (41.9)	—
8α-*N*1-Histidyl riboflavin[d,e]	445 (12.0)	355 (9.3)	268 (35.3)	221 (37.4)	5.75
8α-*N*3-Histidyl riboflavin[f]	445 (12.0)	355 (8.7)	268 (35.3)	221 (37.4)	4.70
Riboflavin[g]	448 (12.3)	372 (10.8)	268 (31.4)	223 (30.1)	

Data are for spectra taken in aqueous buffers at neutral pH. [a]Ref. ***(25)***, [b]Ref. ***(22)***, [c]Ref. ***(24)***, [d]Ref. ***(23)***, [e]Ref. ***(24)***, [f]Ref. ***(24)***, [g]Ref. ***(41)***. nd = not determined.

3.2.4. Characterization of the Aminoacyl Flavin

Various analytical methods have been developed for identifying the chemical nature of the aminoacyl flavin. No doubt modern protein chemical methods such as mass spectrometry would facilitate rapid identification in some circumstances, as has been used for identifying flavin analogs isolated from flavoprotein enzymes [e.g., ***(5)***]. However, the traditional methods of identification are relatively easy to perform and are appropriate in those cases where mass spectrometry facilities are not readily available, or when mass analysis cannot unequivocally identify the aminoacyl flavin (e.g., in determining whether an aminoacyl flavin is of the *N*1- or *N*3 histidyl type).

The 6-*S*-cysteinyl flavins are readily identified by virtue of their highly unusual flavin absorption spectra, which bear no resemblance to the spectra for the 8α methyl-substituted aminoacyl flavins ***(22–25)***. The 6-*S*-cysteinyl flavins have a single absorption peak in the visible region whereas the 8α methyl-substituted aminoacyl flavins are characterized by two peaks in the visible region (**Table 1**). The near-UV peak of the 8α methyl-substituted aminoacyl flavins shows a hypsochromic shift compared with the corresponding peak for riboflavin. Identification of the 8α histidyl flavins can be achieved by investigating fluorescence properties as a function of pH. Flavins with a histidyl moiety at the 8α position exhibit a characteristic increase in fluorescence below pH 7 because at low pH the nonbonding electrons of the nitrogen are not available to interact with the flavin owing to protonation of the nitrogen atom ***(26–28)***. 8α-*N*3-histidyl riboflavin has a pK_a of 4.7 during fluorescence titration ***(29)***, whereas the 8α-*N*1-histidyl flavin has a pK_a of 5.25 ***(23)***. The following method is suitable for investigating fluorescence properties.

1. Prepare equimolar samples of the unknown aminoacyl riboflavin in 100 m*M* citrate-phosphate buffer spanning the pH range 2.5–8.
2. Measure relative fluorescence of each sample at 525 nm following excitation of the sample at 445 nm.
3. Construct a plot of relative fluorescence vs pH. 8α-*O*-tyrosyl and 8α-*S*-cysteinyl riboflavins exhibit a weak pH dependence of fluorescence, whereas the 8α-*N*1-histidyl and 8α-*N*3-histidyl riboflavins display an enhancement in fluorescence of about threefold on moving from pH 7 to pH 3.
4. If an enhancement in fluorescence is seen below pH 7, fit data collected in **step 3** to the following equation to obtain a pK_a value for the aminoacyl riboflavin

$$F = F_L + (F_H \cdot K_a)/(K_a + [\mathrm{H}^+])$$

   where F is the observed relative fluorescence at a given value of pH, F_L is the limiting relative fluorescence at low pH, and F_H is the limiting fluorescence at high pH (*see* **Note 8**).
5. To distinguish further 8α-*N*1-histidyl and 8α-*N*3-histidyl riboflavins, add a small amount of solid sodium borohydride to a solution of the aminoacyl riboflavin (dissolved in 100 m*M* citrate-phosphate buffer, pH 7) and measure fluorescence changes accompanying the addition of the reductant. 8α-*N*1-histidyl riboflavin is reduced by borohydride and reduction is accompanied by a decrease in fluorescence. The 8α-*N*3-histidyl riboflavin isomer is not reduced and there is no fluorescence change on addition of the reductant ***(30)***. Further tests may also be performed to distinguish the two isomers (*see* **Note 9**).

Evidence to support the presence of 8α-*O*-tyrosyl riboflavin can be obtained from analysis of the UV-visible absorption spectrum. In addition to the hypsochromic shift of the near-UV absorption band which is characteristic of all 8α-aminoacyl riboflavins (**Table 1**), the peak maximum at 268 nm is 10–20% higher for 8α-*O*-tyrosyl riboflavin than for the 8α-histidyl riboflavins and 8α-*S*-cysteinyl riboflavin ***(24)***. However, this feature alone should not be taken as proof, since impurities may contribute to absorption in the UV region. Unequivocal identification can be achieved by performing anaerobic acid hydrolysis [16 h at 110°C in 4N methanesulfonic acid containing 0.2% (v/v) tryptamine] of about 5 nmol of the putative 8α-*O*-tyrosyl riboflavin. Determination by amino acid analysis should indicate the presence of approximately equimolar tyrosine *(24)*. Alternatively, fluorescence analysis of the putative 8α-*O*-tyrosyl riboflavin following reductive cleavage to tyrosine and riboflavin can be investigated. The fluorescence of 8α-*O*-tyrosyl riboflavin is only 1.5% that of riboflavin and it is not enhanced following performic acid oxidation (as for 8α-*S*-cysteinyl riboflavin). However, reductive cleavage using the following procedure will lead to enhancement of fluorescence owing to the production of riboflavin.

1. Prepare a sample of the aminoacyl flavin in 0.1 *M* citrate-phosphate buffer, pH 7.0 and measure relative fluorescence at 525 nm following excitation of the sample at 445 nm.
2. Transfer the aminoacyl flavin solution to a glass tonometer and make anaerobic by evacuation and flushing with oxygen-free nitrogen. Add at least 2 mol equivalents of dithionite. Reoxidize the cleaved flavin by admitting air to the tonometer (*see* **Note 10**).
3. Measure relative fluorescence at 525 nm following excitation of the sample at 445 nm. After reductive cleavage and reoxidation, at least a 60-fold increase in relative fluorescence is observed for 8α-*O*-tyrosyl riboflavin.

Enhancement of fluorescence following performic acid oxidation is a convenient method of analyzing for the presence of 8α-*S*-cysteinyl riboflavins. The method for oxidation was described by Walker et al. *(22)* and is adapted here for fluorescence analysis.

1. Prepare a sample of the aminoacyl flavin in 0.1 *M* citrate-phosphate buffer, pH 7.0 and measure relative fluorescence at 525 nm following excitation of the sample at 445 nm.
2. Lyophilize a second sample and redissolve the aminoacyl flavin at 0°C in 0.002 mL of performic acid mixture per nmol of aminoacyl flavin present. Incubate for 2 h at 0°C with occasional shaking and dilute the solution with 5 vol of water at 0°C.
3. Lyophilize the solution and resuspend in 0.1 *M* citrate-phosphate buffer, pH 7.0, and measure relative fluorescence at 525 nm following excitation of the sample at 445 nm. A nine to tenfold increase in relative fluorescence following performic acid oxidation is diagnostic of the presence of 8α-*S*-cysteinyl riboflavin.

Additional evidence for the presence of 8α-*S*-cysteinyl riboflavin can be obtained by acid treatment of the aminoacyl riboflavin to produce free cysteine, which is then detected with chloroplatinic acid *(31)*.

Oxidation with performic acid (to produce the sulfone) followed by acid treatment yields a negative chloroplatinate test (*see* **Note 11**).

Finally, it is important to note that routes to the synthesis of 8α aminoacyl riboflavins *(24,26,32)* and 6-*S*-cysteinyl riboflavin *(33)* have been reported.

Unequivocal identification of aminoacyl flavins can therefore also be made by comparing the mobilities of unknown samples with synthetic standards during thin-layer chromatography, high-performance liquid chromatography or by mass analysis.

4. Notes

1. Thioether-linked flavins require prior treatment with performic acid to overcome the internal quenching of fluorescence for these flavin derivatives. Treatment of 8α-*S*-cysteinyl flavins with performic acid converts the thioether into the corresponding sulfone [8α-*S*-cysteinylsulfonyl flavin ***(34)***] and quenching is lost since the sulfone no longer has nonbonding orbitals to interact with the flavin system ***(35)***.
2. By far the most common procedures used for the identification of covalent flavoproteins have relied on the use of trypsin and chymotrypsin for fragmentation of the enzyme. The methods used are similar to the one described in this chapter, which uses pepsin to fragment the protein.
3. The use of phosphocellulose and 5% formic acid is particularly effective for this purification step. At acid pH, most peptides remain bound to phosphocellulose due to the presence of the α-amino group. Flavin peptides that lack basic amino acyl residues do not bind to the support due to the presence of the negatively charged phosphate group on the flavinylated peptide. A final "polishing" step may be required [as for *Chlorobium* cytochrome c_{553} ***(13)***], but in the majority of cases the isolated flavinylated peptide will be sufficiently pure after phosphocellulose chromatography to enable further analysis.
4. Release of AMP at this stage indicates that the flavinylated peptide contains a FAD derivative. Many methods exist for analysis of AMP, but a simple procedure involves ion-pair reverse-phase HPLC ***(6)*** using a 3 μm ODS-Hypersil column (2.1. mm × 10 cm) (Watford, Hertfordshire, UK) and a triethylamine ion-pair/methanol gradient ***(36)***. Following application of a sample of the digest to the column, elute with a linear gradient from 83.3 m*M* triethylamine buffer, pH 6.0, methanol (98:2) to 83.3 m*M* triethylamine buffer, pH 6.0, methanol (90:10) at a flow rate of 0.2 mL/min. Monitor nucleotide elution at 258 nm, or if available, by diode array. Use samples of authentic AMP as standards. If HPLC facilities are not available, release of AMP can be detected by poly(ethyleneimine)-cellulose thin-layer chromatography ***(37)***.
5. Some workers have employed acid hydrolysis (6N HCl at 95°C for 16 h *in vacuo*) to generate the aminoacyl flavin, rather than, or as a supplement to, using aminopeptidase *M* [e.g., ***(9)*** for 8α-*N*(3)-histidyl riboflavin; ***(10)*** for 8α-*N*(1)-histidyl riboflavin; ***(13)*** for 8α-*S*-cysteinyl riboflavin; (***20***) for combined use with aminopeptidase *M*]. However, care must be exercised when using this approach since mixed products are often obtained due to dehydration of the ribityl chain to form 2', 5'-anhydroriboflavin derivatives when exposed to high temperatures and acid conditions ***(30,38)***, and in the case of 8α-*S*-cysteinyl riboflavin 8-formylriboflavin is a major product of acid hydrolysis ***(34)***. Identification of the acid

hydrolysis products usually requires comparison with authentic standards during high-voltage paper electrophoresis [e.g., ***(10,34)***].

6. Depending on the volume of the aminoacyl flavin solution, desalting may be advisable prior to lyophilization.
7. If desired, the purity of the fractions can be checked by performing analytical thin-layer chromatography using the following solvent system: isopropyl alcohol/water (3:1 v/v). Other purification methods may also be suitable e.g., ion-exchange chromatography or HPLC.
8. An example of this method has been reported ***(20)*** to distinguish between 8α-*N*(1)-histidyl and 8α-*N*(3)-histidyl riboflavin following the preparation of an unknown aminoacyl riboflavin from bacterial sarcosine oxidase.
9. Further tests to distinguish between the two isomers have been devised. In brief, the isomers can be distinguished following treatment with sodium borohydride, methyl iodide and prolonged exposure to acid to yield N1 and N3 methylhistidine ***(12)***. Alternatively, the two isomers can be distinguished by paper electrophoresis at pH 5.0 or isoelectric focusing ***(39)***].
10. Reductive cleavage requires 1 mol equivalent of dithionite; the second mol equivalent reduces the flavin to the dihydroflavin form. If sufficient material is available, reduction can be monitored by absorption spectroscopy. Bleaching of the absorbance at 446 nm occurs on addition of the second mol equivalent of dithionite. Following reoxidation of the material after admitting air to the tonometer, the absorption spectrum resembles that of riboflavin rather than 8α-*O*-tyrosyl riboflavin i.e., there is no hypsochromic shift of the near UV peak ***(24)***.
11. If further analysis is required, experimental details for the chloroplatinate test are published ***(40)***.

References

1. Mewies, M., McIntire, W. S., and Scrutton, N. S. (1998) Covalent attachment of flavin adenine dinucleotide (FAD) and flavin mononucleotide (FMN) to enzymes: the current state of affairs. *Protein Sci.* **7,** 1–14.
2. Edmondson, D. E. and De Francisco, R. (1992) Structure, synthesis and physical properties of covalently bound flavins and 6- and 8-hydroxyflavins, in *Chemistry and Biochemistry of Flavoenzymes*, vol. 1 (Muller, F., ed.), Boca Raton, FL, CRC Press, pp. 73–103.
3. Williamson, G. and Edmondson, D. E. (1985) Effects of pH on oxidation-reduction potentials of 8α-*N*-substituted flavins. *Biochemistry* **24,** 7790–7797.
4. Kim, J., Fuller, J. H., Kuusk, V., Cunane, L., Chen, Z., Mathews, F. S., and McIntire, W. S. (1995) The cytochrome subunit is necessary for covalent FAD attachment to the flavoprotein subunit of *p*-cresol methylhydroxylase. *J. Biol. Chem.* **263,** 31,202–31,209.
5. Mewies, M., Basran, J., Packman, L. C., Hille, R., and Scrutton, N. S. (1997) Involvement of a flavin iminoquinone methide in the formation of 6-hydroxy FMN in trimethylamine dehydrogenase: a rationale for the existence of 8α-methyl and C6-linked covalent flavoproteins. *Biochemistry* **36,** 7162–7168.

6. Scrutton, N. S., Packman, L. C., Mathews, F. S., Rohlfs, R. J., and Hille, R. (1994) Assembly of redox centers in the trimethylamine dehydrogenase of bacterium W3A1: properties of the wild-type enzyme and a C30A mutant expressed from a cloned gene in *Escherichia coli. J. Biol. Chem.* **269,** 13,942–13,950.
7. Nagursky, H., Bichler, V., and Brandsch, R. (1988) Phosphoenolpyruvate-dependent flavinylation of 6-hydroxy-D-nicotine oxidase. *Eur. J. Biochem.* **177,** 319–325.
8. Packman, L. C., Mewies, M., and Scrutton, N. S. (1995) The flavinylation reaction of trimethylamine dehydrogenase: analysis by directed mutagenesis and electrospray mass spectrometry. *J. Biol. Chem.* **270,** 13,186–13,191.
9. Ohishi, N. and Yagi, K. (1979) Covalently bound flavin as prosthetic group of choline oxidase. *Biochem. Biophys. Res. Comm.* **86,** 1084–1088.
10. Kiuchi, K., Nishikimi, M., and Yagi, K. (1982) Purification and characterisation of L-gulonolactone oxidase from chicken kidney microsomes. *Biochemistry* **21,** 5076–5082.
11. Kutchan, T. M. and Dittrich, H. (1995) Characterisation and mechanism of the berberine bridge enzyme, a covalently flavinylated oxidase of benzophenanthridine alkaloid biosynthesis in plants. *J. Biol. Chem.* **270,** 24,475–24,481.
12. Singer, T. P. and McIntire, W. S. (1984) Structure, properties and determination of covalently bound flavins. *Methods Enzymol.* **66,** 253–264.
13. Kenney, W. C., McIntire, W. S., and Yamanaka, T. (1977) Structure of the covalently bound flavin of *Chlorobium* cytochrome c553. *Biochim. Biophys. Acta* **483,** 467–474.
14. Bruhmuller, M., and Decker, K. (1973) Covalently bound flavin in D-6-hydroxynicotine oxidase from *Arthrobacter oxidans. Eur. J. Biochem.* **37,** 256–258.
15. Kenney, W. C., Edmondson, D. E., and Seng, R. L. (1976) Identification of the covalently bound flavin of thiamin dehydrogenase. *J. Biol. Chem.* **251,** 5386–5390.
16. Kenney, W. C., Edmondson, D. E., Singer, T. P., Nishikimi, M., Noguchi, E., and Yagi, K. (1979) Identification of the covalently-bound flavin of L-galactonolactone oxidase from yeast. *FEBS Lett.* **97,** 40–42.
17. Kenney, W. C., Singer, T. P., Fukuyama, M., and Miyake, Y. (1979) Identification of the covalently bound flavin prosthetic group of cholesterol oxidase. *J. Biol. Chem.* **254,** 4689–4690.
18. Steenkamp, D. J., Kenney, W. C., and Singer, T. P. (1978) A novel type of covalently bound coenzyme in trimethylamine dehydrogenase. *J. Biol. Chem.* **253,** 2812–2817.
19. McIntire, W., Edmondson, D. E., Singer, T. P., and Hopper, D. J. (1980) 8α-*O*-tyrosyl FAD: a new form of covalently bound flavin from *p*-cresol methylhydroxylase. *J. Biol. Chem.* **255,** 6553–6555.
20. Kvalnes-Krick, K. and Jorns, M. S. (1986) Bacterial sarcosine oxidase: comparison of two multisubunit enzymes containing both covalent and noncovalent flavin. *Biochemistry* **25,** 6061–6069.
21. De Jong, E., van Berkel, W. J. H., van der Zwan, R. P., and de Bont, J. A. M. (1992) Purification and characterisation of vanillyl-alcohol oxidase from *Penicillium simplicissium. Eur. J. Biochem.* **208,** 651–657.

22. Walker, W. H., Kearney, E. B., Seng, R. L., and Singer, T. P. (1971) The covalently-bound flavin of hepatic monoamine oxidase: identification and properties of cysteinyl riboflavin. *Eur. J. Biochem.* **24,** 328–331.
23. Edmondson, D. E., Kenney, W. C., and Singer, T. P. (1976) Structural elucidation and properties of 8α-(N^1-histidyl) riboflavin: the flavin component of thiamine dehydrogenase and cyclopiazonate oxidocyclase. *Biochemistry* **15,** 2937–2945.
24. McIntire, W., Edmondson, D. E., Hopper, D. J., and Singer, T. P. (1981) 8α-(*O*-Tyrosyl)flavin adenine dinucleotide, the prosthetic group of bacterial *p*-cresol methylhydroxylase. *Biochemistry* **20,** 3068–3075.
25. Steenkamp, D. J., McIntire, W., and Kenney, W. C. (1978). Structure of the covalently bound coenzyme of trimethylamine dehydrogenase: evidence for a 6-substituted flavin. *J. Biol. Chem.* **253,** 2818–2824.
26. Singer, T. P., and Edmondson, D. E. (1980) Structure, properties, and determination of covalently bound flavins. *Methods Enzymol.* **66,** 253–264.
27. Salach, J., Walker, W. H., Singer, T. P., Ehrenberg, A., Hemmerich, P., Ghisla, S., and Hartman, U. (1972) Studies on succinate dehydrogenase: site of attachment of the covalently-bound flavin to the peptide chain. *Eur. J. Biochem.* **26,** 267–278.
28. Kenney, W. C. and Walker, W. H. (1972) Synthesis and properties of 8α-substituted riboflavins of biological importance. *FEBS Lett.* **20,** 297–301.
29. Walker, W. H., Singer, T. P., Ghisla, S., and Hemmerich, P. (1972) Studies on succinate dehydrogenase: 8α-histidyl FAD as the active center of succinate dehydrogenase. *Eur. J. Biochem.* **26,** 279–289.
30. Edmondson, D. E. (1976) 2',5'-Anhydro-8α-histidyl flavins: their occurrence and structural elucidation. *Fed. Proc.* **35,** 1542.
31. Kearney, E. B., Salach, J. I., Walker, W. H., Seng, R. L., Kenney, W., Zeszotek, E., and Singer, T. P. (1971) The covalently-bound flavin of hepatic monoamine oxidase. 1. Isolation and sequence of a flavin peptide and evidence for binding at the 8α position. *Eur. J. Biochem.* **24,** 321–327.
32. Edmondson, D. E., Kenney, W. C., and Singer, T. P. (1978) Synthesis and isolation of 8α-substituted flavins and flavin peptides. *Methods Enzymol.* **53,** 449–465.
33. Ghisla, S., Kenney, W. C., Knappe, W. R., McIntire, W. S., and Singer, T. P. (1980) Chemical synthesis and some properties of 6-substituted flavins. *Biochemistry* **19,** 2537–2544.
34. Kenny, W. C. and Singer, T. P. (1977) Evidence for a thioether linkage between the flavin and polypeptide chain of *Chromatium* cyctochrome c_{552}. *J. Biol. Chem.* **252,** 4767–4772.
35. Falk, M. C. and McCormick, D. B. (1976) Synthetic flavinyl peptides related to the active site of mitochondrial monoamine oxidase II. Fluorescence properties. *Biochemistry* **15,** 646–653.
36. Perrett, D., Bhuste, L., Patel, J., and Herbert, K. (1991) Comparative performance of ion-exchange and ion-paired reversed phase high-performance liquid-chromatography for the determination of nucleotides in biological samples. *Biomed. Chromatogr.* **5,** 207–211.

37. Randerath, K. and Randerath, E. (1967) Thin-layer separation methods for nucleic acid derivatives. *Methods Enzymol.* **12,** 323–347.
38. Kenney, W. C., Edmondson, D. E., Singer, T. P., Steenkamp, D. J., and Schabort, J. C. (1976) Identification and properties of the covalently bound flavin of β-cyclopiazonate oxidocyclase. *Biochemistry* **15,** 4931–4935.
39. McIntire, W. S., Singer, T. P., Ameyama, M., Adachi, O., Matsushita, K., and Shinagawa, E. (1985) Identification of the covalently-bound flavins of D-gluconate dehydrogenase from *Pseudomonas aeroginosa* and *Pseudomonas fluorescens* and of 2-keto-D-gluconate dehydrogenase from *Gluconobacter melanogenus. Biochem. J.* **231,** 651–654.
40. Smith, I. (1960) Thin layer chromatography, in *Chromatographic and Electrophoretic Techniques,* vol. 1 (Smith, I., ed.), 2nd ed., Interscience, New York, pp. 66–103.
41. Koziol, J. (1971) Fluorometric analyses of riboflavin and its coenzymes. *Methods Enzymol.* **18B,** 253–285.

14

Flavin Synthesis and Incorporation into Synthetic Peptides

R. Eryl Sharp and P. Leslie Dutton

1. Introduction

There exists a hierarchical approach to the study of flavoprotein function in biology: investigation in natural systems (by far the most common) *(1)*; using semisynthetic systems, pioneered by Kaiser *(2,3)*, where the flavin is attached near to the active site of a natural enzyme and the flavin chemistry modifies the catalytic behavior of the enzyme; and in fully synthetic systems where the flavin is attached to novel chemical frameworks *(4)* or synthetic peptides *(5–7)*. The focus of our research falls into the last category, as studies aimed at understanding the molecular basis of catalytic fitness in enzymes are often confounded by the sheer complexity of natural proteins. Using insights drawn from natural systems, we design and synthesize minimalistic structures that assemble the component peptides and prosthetic groups into a functional array. We term these working structures *molecular maquettes*. The peptide framework of choice is the well-characterized, robust, and structurally defined tetra α-helix bundle, which is a dimer of helix-loop-helix secondary structure motifs, that pack together to form a well defined hydrophobic core *(8)*. To date, there has been much success with this approach, notably the incorporation of heme into tetra α-helix bundle proteins, via *bis*-histidine axial coordination to the porphyrin iron *(9)*.

From a thorough investigation of existing flavoprotein structures within the Brookhaven database, we concluded that engineering a flavin binding site into our existing maquette framework is premature and would pose considerable design problems. It is evident that flavoproteins, including even the simple flavodoxins, possess a completely different architecture to that offered by tetra-α-helix-bundles *(10)*. However, many flavoproteins contain flavin covalently

From: Methods in Molecular Biology, vol. 131: Flavoprotein Protocols
Edited by: S. K. Chapman and G. A. Reid © Humana Press Inc., Totowa, NJ

bound to the polypeptide and the point of attachment can occur at a variety of positions around the isoalloxazine ring ***(11)*** *see* Chapter 13. Therefore, there exists a precedent for localizing the flavin within the helix-loop-helix framework by covalently attaching it to an amino acid side chain situated within the hydrophobic core of the bundle.

The maquette design was similar to that of our earlier peptides ***(9)***, in that it comprised two 62-amino acid helix-loop-helix subunits. Each subunit contains a single cysteine to which flavin (7-acetyl-10-methyl-isoalloxazine) is covalently attached and two histidines appropriately positioned for *bis*-his coordination of heme cofactors. Both flavins and hemes are situated within the hydrophobic core of the protein. This minimalistic designed flavo-heme protein is able to support light-induced intramolecular electron transfer between the flavin and heme, which will ultimately lead to the description of the parameters that govern the nature of this and related processes ***(7)***.

In this chapter we describe the synthesis of a nonnatural flavin (which nevertheless exhibits all the salient physical chemical properties of natural flavins) with the appropriate functionality for facilitating attachment to a reactive thiol group on a cysteine in a synthetic peptide. The methodology we report for synthesis and subsequent purification of the flavinated peptide is straightforward, simple in execution, and can readily be applied for attachment of a variety of flavin derivatives ***(12)*** to any peptide system with a single reactive thiol species.

2. Materials

All chemicals used are reagent grade and are obtained from Sigma (St. Louis, MO)/Aldrich (Milwaukee, WI), unless otherwise indicated.

2.1. Synthesis of 7α-Bromo-Acetyl-10-Methyl-Isoalloxazine

2.1.1. Step 1

1. *para*-Chloro-acetophenone.
2. Concentrated sulfuric acid.
3. Concentrated nitric acid.
4. Distilled water.
5. Methanol (Burdick and Jackson, Muskegon, MI).

2.1.2. Step 2.

1. Product of **step 1** (4-chloro-3-nitro-acetophenone), store at 4°C in the dark.
2. Ethanol.
3. 40% Methylamine in distilled water.
4. 4 *M* Ammonium hydroxide.
5. Methanol.

2.1.3. Step 3.

1. Product of **step 2** (4-methylamino-3-nitroacetophenone), store at 4°C in the dark.
2. Ethyl acetate.
3. Platinum IV oxide (Adams catalyst).
4. Hydrogenation apparatus (**Fig. 2**).
5. Kaiser test: 0.1 g mL^{-1} Ninhydrin in ethanol, 4 g mL^{-1} phenol in ethanol, 20 µ*M* potassium cyanide in pyridine.

2.1.4. Step 4.

1. Product of **step 3** (7-acetyl-10-methyl-isoalloxazine), store at 4°C in the dark.
2. Glacial acetic acid.
3. Ether.

2.2. Peptide Flavination

1. Pure peptide as a dry, lyophilized powder, stored at –20°C.
2. 0.1 *M* Tris-HCl buffer at pH 7.5.
3. Dimethylformamide.
4. 7α-Bromo-acetyl-10-methyl-isoalloxazine.

2.3. Equipment

Reflux apparatus, hydrogenation apparatus depicted in **Fig. 1**, vacuum freeze dryer for lyophilization, thin layer chromatography, UV-visible spectrophotometer (Lambda 2, Perkin-Elmer, Norwalk, CT). C_{18} (Vydac columns, Vydac, Hesperia, CA) analytical and preparatory high-pressure liquid chromatography (HPLC) (System Gold, Beckman, Fullerton, CA). The following solvents are used for HPLC (*see* **Note 1**): 0.1% (v/v) trifluroacetic acid in Millipore filtered water (Millipore, Bedford, MA) (>17 MΩ resistance), 0.1% (v/v) trifluroacetic acid in HPLC-grade acetonitrile (Burdick and Jackson). Laser desorption mass spectrometer facility.

3. Method

3.1. Synthesis of 7α-Bromo-Acetyl-10-Methyl-Isoalloxazine

The salient features of this procedure are illustrated schematically in **Fig. 2**. The protocol is essentially that described by Levine and Kaiser *(13)*, but with certain modifications at **steps 3** and **4**.

1. Preparation of 4-chloro-3-nitro-acetophenone. To a stirred solution of 80 mL concentrated sulfuric acid and 20 mL concentrated nitric acid in an ice-salt bath is added 10.4 *g* (0.068 mol) of *p*-chloro-acetophenone dropwise over a period of 30 min. The solution is stirred for an additional 15 min and then poured over 0.5 L of distilled ice water. The resulting crystals are collected, washed with ice water

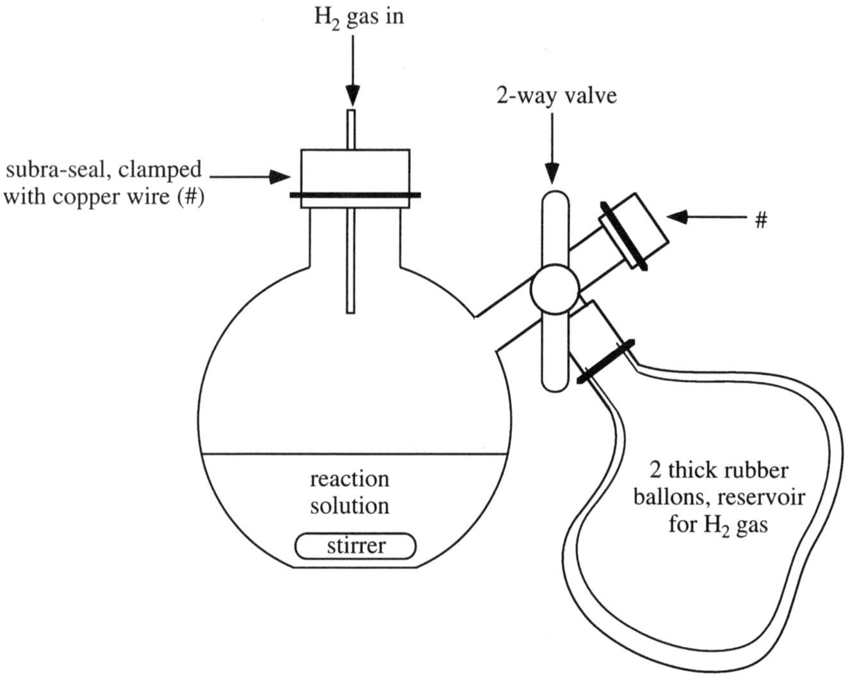

Fig. 1. Diagram of apparatus used for hydrogenation of 4-methyl-3-nitro-acetophenone. Experimental detail is given in **Subheadings 3.1.3.** and **4.2.**

several times and recrystallized from methanol (by dissolving the crystals in the minimum possible volume of hot methanol and allowing the solution to cool slowly). The yield/ recovery is typically >90%.

2. Preparation of 4-methylamino-3-nitro-acetophenone. To a stirred solution containing 2.0 *g* (0.01 mol) of 4-chloro-nitro-acetophenone in 15 mL of ethanol is added 4.0 *g* (0.057 mol) of a 40% solution of methylamine in water. After refluxing for 3 h, 25 mL of 4 *M* ammonium hydroxide is added and the turbid solution is allowed to stand at 4°C for several hours. The precipitated product is collected and recrystallized from methanol. Typical yield/recovery is 75%.
3. Preparation of 7-acetyl-10-methyl-isoalloxazine. 502.1 mg (2.59 mmol) of 4-methyl-3-nitro-acetophenone is dissolved in 150 mL of ethyl acetate and catalytically hydrogentated over 125 mg of Adams catalyst. The reaction is performed in the apparatus illustrated in **Fig. 1** (*see* **Note 2**). Hydrogenation is typically complete within 2 h and could be determined by eye, since the color of the solution changes from green to clear. Thin layer chromatography can also be used to monitor the reaction progress, using 5% methanol, 95% chloroform as solvents, with silica gel as the chromatographic material. The reactants and products are identified by the Kaiser test ***(14)*** (*see* **Note 3**). The appearance of the gel is a spot-

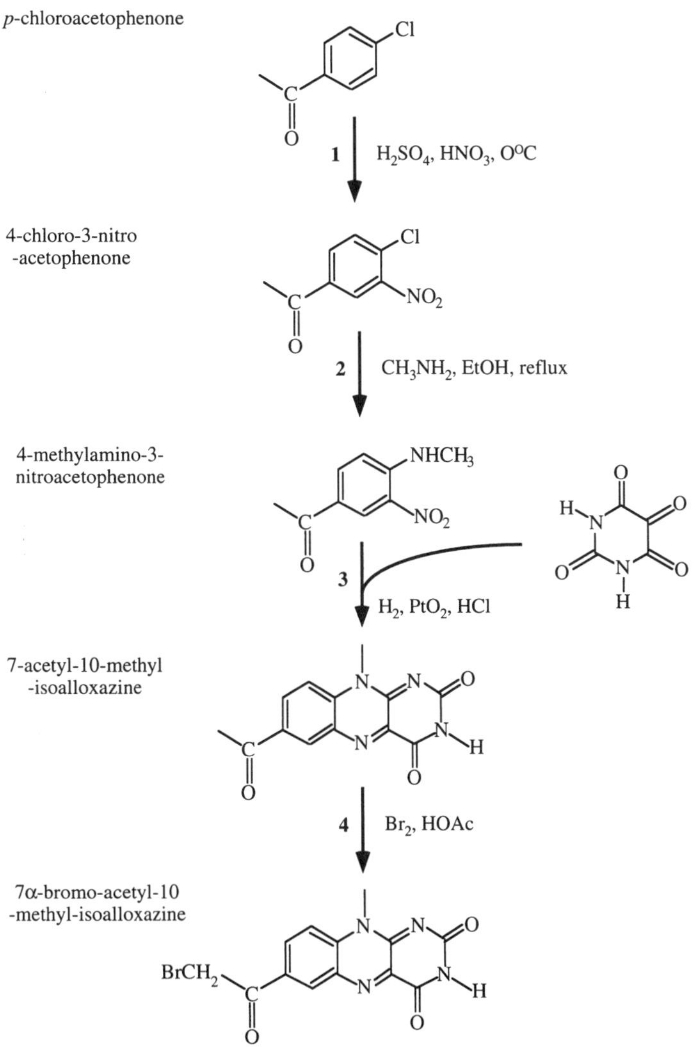

Fig. 2. Flowchart illustrating 7α-bromo-acetyl-10-methyl-isoalloxazine synthesis. The experimental detail is described in methods **Subheading 3.1.**

to-spot reaction, i.e., there are no intermediates. The colorless solution is filtered to remove the catalyst and evaporated to dryness using a rotor evaporator. The obtained residue is dissolved in 100 mL of absolute ethanol and 417.4 mg (2.61 mmol) of alloxan monohydrate suspended in 5 mL concentrated hydrochloric acid is added. The mixture is refluxed for 15 min. The product which crystallizes upon cooling to 4°C is collected and washed with cold ether. The yield/recovery is typically 75% (*see* **Note 4**). The purity of the product is confirmed by analytical HPLC (**Fig. 3A**) and the molecular weight determined by laser desorption mass spectrometry (**Fig. 4A**).

4. Preparation of 7α-bromo-acetyl-10-methyl-isoalloxazine. 5 mL of 0.26 *M* bromine (1.3 mmol) in glacial acetic acid is added dropwise over a period of 5 min to a warm solution (80–90°C) of 292.6 mg (1.08 mmol) 7-acetyl-10-methyl-isoalloxazine in 35 mL of glacial acetic acid. The solution is stirred for an additional 10 min and then cooled to room temperature. Towards the end of the reaction the product starts to precipitate and this is further enhanced by cooling the solution to 4°C. The precipitate is collected and washed with cold ether and then dried overnight under high vacuum. The composition of the final mixture is determined by HPLC and the predominant presence of 7α-bromo-acetyl-10-methyl-isoalloxazine was confirmed by mass spectrometry (**Figs. 3A** and **4A**). Pure 7α-bromo-acetyl-10-methyl-isoalloxazine could be obtained by subjecting the heterogeneous sample to preparatory HPLC (*see* **Note 5**). 10 mg of the sample is dissolved in 6 mL of 0.1% trifluroacetic acid/water, sonicated for 5 min and then centrifuged at 4000 rpm to pellet any undissolved material (the solubility is quite low and several additions may be required to fully dissolve the sample). Five milliliters of the sample is injected onto the HPLC column which is run under exactly the same conditions as the analytical HPLC, except that a flow rate of 10 mL min^{-1} was used (**Fig. 3A**). The peak corresponding to 7α-bromo-acetyl-10-methyl-isoalloxazine, which elutes at about 20 min, is collected. The trifluroacetic acid and acetonitrile are evaporated by heating the sample slightly with a hot air drier and bubbling with nitrogen. The sample is frozen and freeze-dried to remove the remaining solvents. The purity of the sample is confirmed by analytical HPLC and the molecular weight by laser desorption mass spectrometry.

3.2. Peptide Synthesis

The chemistry and methodology of solid phase peptide synthesis has been well described elsewhere and will not be reiterated here ***(15)***.

1. Peptide Design: In our present study for synthesis of flavoprotein *maquettes,* we construct 62 amino-acid-length peptides. The designed helix-loop-helix peptide (α_2) has the sequence Ac-L•KKLREEA•LKLLEEF•KKLLEEH•LKWLE GGGG-GGGGELLKL• HEELLKK•CEELLKL•AEERLKK•L-$CONH_2$, where Ac is the *N*-terminal acetyl group and the dots emphasize the underlying heptad repeat pattern of the helices. The cysteine at position 48 to which the flavin is attached is highlighted in the sequence.
2. Peptide synthesis is performed on a continuous-flow Milligen 9050 solid phase synthesizer (Millipore) using standard $F_{moc}/^{t}Bu$ methodology ***(15)***. The yield of total peptide cleaved from the resin is about 95% and the subsequent yield of the desired reverse-phase preparatory HPLC-purified product is 15% of the total peptide.

Fig. 3. HPLC elution profiles of: (A) 7-acetyl-10-methyl-isoalloxazine and 7α-bromo-acetyl-10-methyl-isoalloxazine. The solvents used were acetonitrile and Millipore fil-

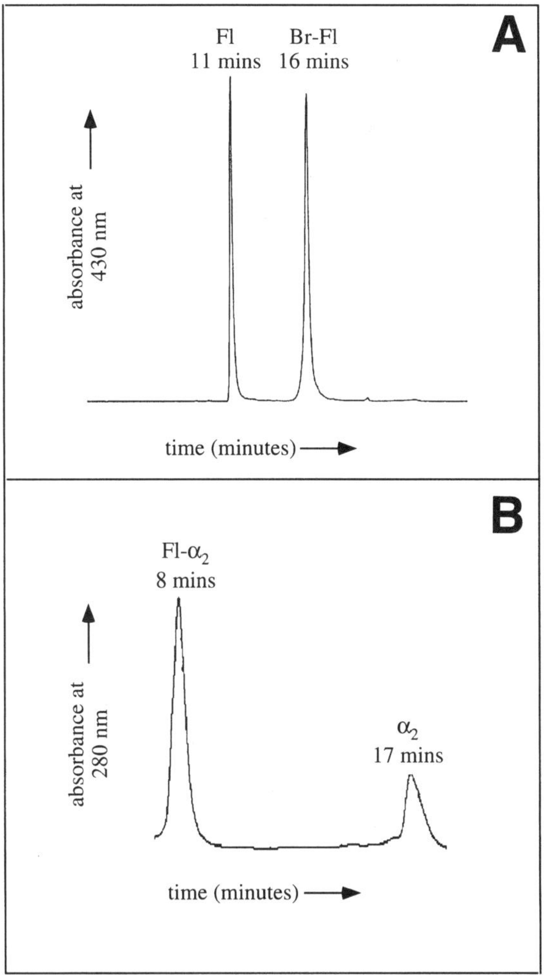

tered water (Millipore) (both containing 0.1% v/v trifluroacetic acid). Initial column equilibration was at a solvent ratio of 5% acetonitrile and 95% water, flow rate 1 mL min^{-1}. The two flavin species were resolved by running a gradient of 5–25% acetonitrile over 25 min at a flow rate of 1 mL min^{-1}, under which conditions the elution times of flavin and monobrominated flavin are well resolved at 11 and 16 min, respectively. Sample detection was performed at 430 nm, at which wavelength the flavin species absorb light stongly. (B) Peptide (α_2) and flavinated-peptide (Fl-α_2). The same solvents were used, with initial equilibration at 45% acetonitrile and 55% water, flow rate 1 mL min^{-1}. The two species were resolved by running a gradient of 45 to 52% acetonitrile over 28 min at a flow rate of 1 mL min^{-1}, under which conditions the elution times of Fl-α_2 and α_2 were well resolved at 8 and 17 min, respectively. Sample detection was performed at 280 nm, at which wavelength both the flavin and peptide tryptophan absorb light.

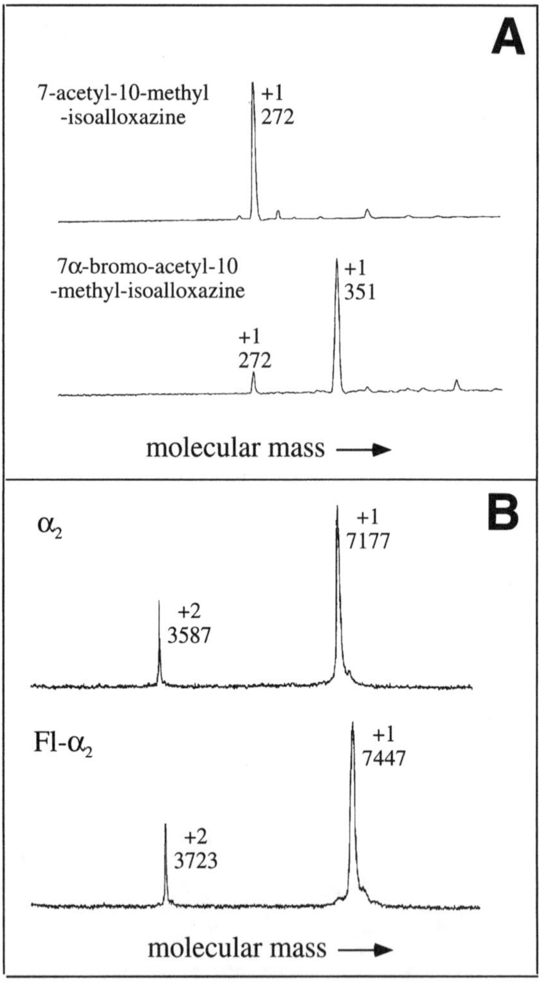

Fig. 4. Laser desorption mass spectrometry of: (A) 7-acetyl-10-methyl-isoalloxazine and 7α-bromo-acetyl-10-methyl-isoalloxazine. The determined molecular mass to charge ratio are reported beside the major peaks, which correspond to the singly ionized flavin species in the upper spectrum and a comparison of the reaction products from methods **Subheading 3.1.4.**, in the lower spectrum. The smaller amplitude peaks are from ionized matrix material and are not due to sample contamination. (B) Peptide (α_2) and flavinated-peptide (Fl-α_2). The singly and doubly ionized species are shown in the figure. In all cases, the experimentally determined molecular masses agree exactly with the calculated values.

3.3. Peptide Flavination

The covalent attachment of the flavin to the peptide cysteine is illustrated schematically in **Fig. 5**.

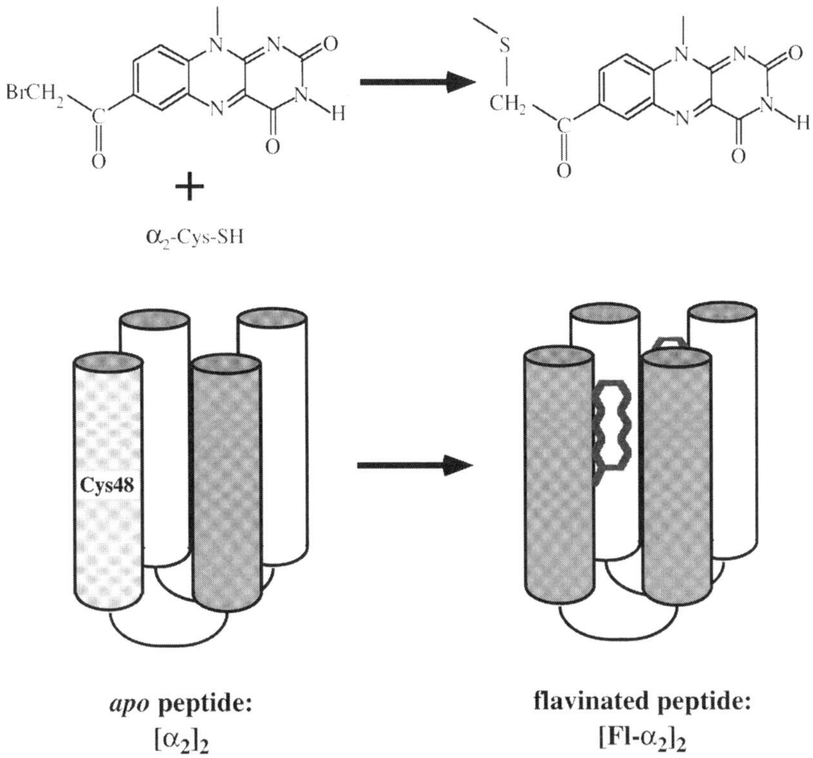

Fig. 5. Scheme illustrating covalent attachment of 7α-bromo-acetyl-10-methyl-isoalloxazine to peptide. The experimental detail is described in **Subheading 3.3.**

1. The desired amount of freeze-dried peptide (α_2) is dissolved in about 2 mL of 0.1 *M* Tris·HCl buffer at pH 7.5 (typically about 2 µmoles peptide).
2. The concentration is determined spectrophotometrically from the UV absorbance spectrum of tryptophan (Trp) ($\varepsilon_{280nm} = 5,700\ M^{-1}cm^{-1}$), all the peptides we use contain one Trp in their sequence.
3. 7α-bromo-acetyl-10-methyl-isoalloxazine is dissolved in dimethylformamide and its concentration determined spectrophotometrically from the blue visible absorption of the flavin ($\varepsilon_{427nm} = 10,900\ M^{-1}cm^{-1}$ in aqueous buffer.
4. A fivefold molar excess of 7α-bromo-acetyl-10-methyl-isoalloxazine is added to the peptide solution and the buffer:DMF ratio adjusted to 1:1 (v/v).
5. The flavination reaction is performed at room temperature with stirring in the dark. After about 5 h the reaction is >95% complete as judged by analytical HPLC (**Fig. 3B**). The progress of the reaction can be conveniently monitored by periodically analyzing aliquots of the reaction using analytical HPLC (**Fig. 3B**).

6. After flavination the solution is dialyzed (using 3500 Dalton molecular weight cut-off tubing) overnight against 2 L of distilled water at 4°C in the dark. This removes nearly all of the DMF and unreacted 7α-bromo-acetyl-10-methyl-isoalloxazine.
7. The dialyzed sample is then concentrated using centriprep 3000 Dalton molecular mass cutoff concentrators (Amicon, Beverly, MA) until the volume is about 5 mL.
8. The flavinated peptide (Fl-α_2) is purified to homogeneity by preparatory HPLC. The column is preequilibrated with 43% acetonitrile, 57% water at a flow rate of 10 mL min^{-1} and a gradient of 43–48% acetonitrile over 30 min is run. The Fl-α_2 elutes after about 15 min, appropriate fractions are combined and the trifluroacetic acid and acetonitrile evaporated by heating the sample slightly with a hot air drier and bubbling with nitrogen.
9. The sample is frozen and freeze-dried to remove the remaining solvents.
10. Analytical HPLC and laser desorption mass spectrometry (**Figs. 3B** and **4B**) are used to confirm the purity of the flavinated peptide and the molecular mass of the flavinated peptide.

4. Notes

1. The 0.1% trifluroacetic acid lowers the pH of the HPLC solvents to 2.0, which maintains the cysteine thiol in the peptide in the fully protonated state, preventing disulfide bond formation.
2. Hydrogen gas is flammable and potentially explosive when it comes into contact with air, therefore the hydrogenation reaction is potentially dangerous and appropriate safety measures should be adopted. The apparatus depicted in **Fig. 2** is assembled in a fume hood along with a small hydrogen gas cylinder, connected to each other by a short section of copper tubing to minimize gas exchange. The fume hood is closed and about 2 L of hydrogen gas is passed into the balloons; once filled, the hydrogen gas source is isolated and the experiment initiated. When the reaction is complete, any remaining hydrogen is slowly vented into the operating fume hood. At such a low partial pressure of hydrogen, there is little danger of ignition. The operator should wear appropriate safety clothing.
3. Kaiser test: All three stock reagents are mixed in a 1:1:1 ratio (ninhydrin in ethanol:phenol in ethanol and potassium cyanide in pyridine) to a volume of about 0.5 mL. The test solution is applied dropwise to the dried silica gel upon which the hydrogenation reaction mixture has been run. Then the gel is again dried by heating gently with a hot air dryer and examined for reaction. A blue spot at the top of the solvent front indicates the presence of the product and a yellow spot at the beginning of the solvent front, the reactant.
4. The literature preparation suggests recrystallization of the product from dimethylsulfoxide. This is difficult to do and in our experience the product is pure enough to use without subjecting it to this step.
5. The literature preparation states that at the end of the reaction there should be less than 5% starting material present and no detectable amount of dibrominated species. In our experience there is always some unbrominated species remaining, typically about 25%, but very little dibrominated species (<2%), with the major-

ity of the products being the desired monobrominated species. However, if the temperature is too high (>90°C) and the reaction is allowed to proceed for longer than 15 min, then a substantial amount of the dibrominated species accumulates.

Acknowledgments

Mass spectrometry analyses were performed by the University of Pennsylvania Protein Chemistry Laboratory. This work was supported by the National Institutes of Health Grant GM 41048 and in part by the National Science Foundation. R.E.S. gratefully acknowledges receipt of a postdoctoral fellowship from the North Atlantic Treaty Organization.

References

1. Massey, V. (1995) Flavoprotein structure and mechanism. *FASEB J.* **9,** 473–475.
2. Kaiser, E. T. and Lawrence, D. S. (1984) Chemical mutation of enzyme active sites. *Science* **226,** 505–511.
3. Twitchet, M. B., Ferrer, J. C., Siddarth, P., and Mauk, A. G. (1997) Intramolecular electron transfer kinetics of a synthetic flavocytochrome. *J. Am. Chem. Soc.* **119,** 435–436.
4. Seward, E. M., Hopkins, R. B., Sauerer, W., Tam, S.-W., and Diederich, F. (1990) Redox-dependent binding affintiy of a flavin cyclophane in aqueous solution: hydrophobic stacking versus cavity-inclusion complexation. *J. Am. Chem. Soc.* **112,** 1783–1790.
5. Sakamoto, S., Aoyagi, H., Nakashima, N., and Mihara, H. (1996) Design and synthesis of flavin-conjugated peptides and assembly on a gold electrode. *J. Chem. Soc. Perkin Trans.* **2,** 2319–2326.
6. Sharp, R. E., Rabanal, F., and Dutton, P. L. (1996) Design, synthesis and characterisation of a flavocytochrome molecular maquette, in *Flavins and Flavoproteins* (Stephenson, K., ed.), University of Calgary Press, Calgary, Alberta, Canada, pp. 163–166.
7. Sharp R. E., Moser, C. C., Rabanal, F., and Dutton, P. L. (1998) Design, synthesis and characterisation of a photoactivateable flavocytochrome molecular maquette. *Proc. Natl. Acad. Sci. USA* **95,** 10,465–10,470.
8. Bryson, J. W., Betz, S. F., Lu. H. S., Suich, D. J., Zhou, H. X., O'Neil, K. T., and DeGrado, W. F. (1995) Protein design: a hierarchic approach. *Science* **270,** 935.
9. Robertson, D. E., Farid, R. S., Moser, C. C., Urbauer, J. L., Mulholland, S. E., Ravindernath, P., Lear, J. D., Wand, A. J., DeGrado, W. F., and Dutton, P. L. (1994) Design and Synthesis of multi-haem proteins. *Nature* **368,** 425–432.
10. Schulz, G. E. (1992) Binding of nucleotides by proteins. *Curr. Opin. Struct. Biol.* **2,** 61–67.
11. Mewies, M., McIntire, W. S., and Scrutton, N. S. (1998) Covalent attachement of flavin adenine dinucleotide (FAD) and flavin mononucleotide (FMN) to enzymes: the current state of affairs. *Protein Sci.* **7,** 7–20.

12. Slama, J. T., Radziejewski, C., Oruganti, S., and Kaiser, E. T. (1984) Semisynthetic enzymes: charaterization of isomeric flavopapains with widely different catalytic effeiciencies. *J. Am. Chem. Soc.* **106,** 6778–6785.
13. Levine, H. L. and Kaiser, E. T. (1978) Oxidation of dihydronicotinamides by flavopapain. *J. Am. Chem. Soc.* **100,** 7670–7677.
14. Kaiser, E., Colescott, R. L., Bossinger, C. D., and Cook, P. I. (1970) Color test for detection of free terminal amino groups in the solid-phase synthesis of peptides. *Analyt. Biochem.* **34,** 595.
15. Rickwood, D. and Hames, B. D., eds. (1989) *Solid Phase Peptide Synthesis: A Practical Approach,* IRL Press at Oxford University Press, Oxford, UK.

15

Computational Methods in Flavin Research

Lars Ridder, Hans Zuilhof,
Jacques Vervoort, and Ivonne M. C. M. Rietjens

1. Introduction

With the continuously increasing power of computers, quantum chemistry is becoming a valuable theoretical tool in enzyme research. Molecules as large as flavins can now be treated by computational methods of reasonable theoretical level. The present chapter focuses on the possibilities and restrictions of some quantum chemical methods with respect to research on the chemistry of flavin cofactors in enzyme catalysis.

Flavin cofactors are involved in a wide variety of enzymatic reactions like oxidation, reduction and monooxygenation. Consequently, flavin cofactors occur in many different forms including, for example, the oxidized, semiquinone, and reduced states and the C4a-(hydro)peroxyflavin form (**Fig. 1**). In flavoprotein research the geometrical and electronic properties of all these different flavin species are of interest. With the use of quantum chemical methods, many properties, such as optimal geometry, redox potential, charge distribution, dipole moment, and the localization of the electrophilic and nucleophilic reactivity, can be calculated for the different flavin molecules. Furthermore, reactions of flavins with other compounds can be studied and transition states and reaction intermediates can be characterized. These theoretical techniques will become increasingly important to complement experimental results, especially in the case of unstable reaction intermediates, which are difficult to investigate by experimental techniques.

2. Materials

2.1. Software

Quantum chemical computations can be performed with a variety of (often commercially available) programs. Which program is most useful depends on

From: Methods in Molecular Biology, vol. 131: Flavoprotein Protocols
Edited by: S. K. Chapman and G. A. Reid © Humana Press Inc., Totowa, NJ

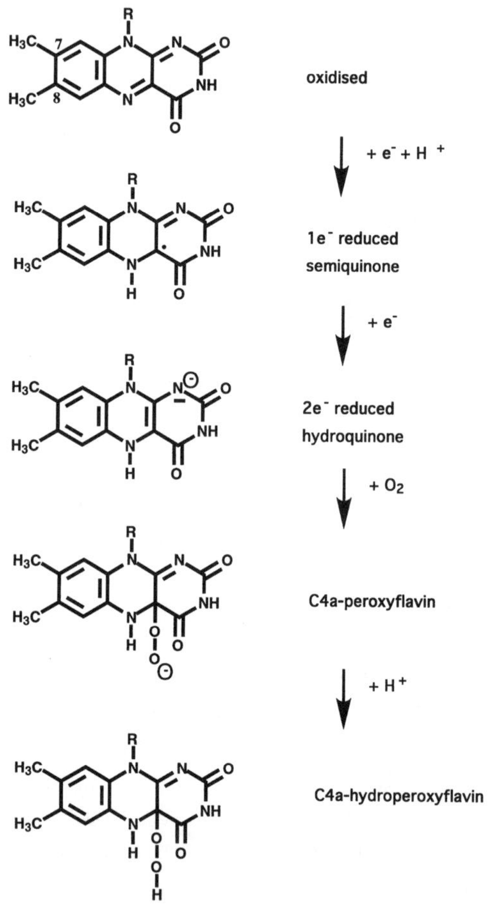

Fig. 1. Various forms of the flavin cofactor relevant for flavin-based biochemical reactions.

the specific goal of the computation, which itself is linked to the quantum chemical method used (vide infra, **Subheading 3.2.**). Reasonably user-friendly programs, e.g., Spartan (for specific information, *see*: http://www.wavefun.com), Hyperchem (http://www.hyper.com), and ChemOffice (http://www.camsoft.com), allow one to draw the structure of the species of interest in a convenient way. Subsequently, the structures can be "cleaned up" via the use of very fast molecular mechanics options (such as MM2 or MMX), which may be desirable for some purposes. With molecular mechanical methods, no electronic information can be derived, but the geometry can be estimated relatively efficiently. In the next step, these programs can be used for the quantum chemical calculations themselves, via simple pull-down menu-options. Within ChemOffice semiempirical methods such as AM1, PM3, and MNDO/d are available, as the full version of this program includes MOPAC97. The same options are also available within Spartan and Hyperchem, which are in many respects similar programs. The graphical interface of ChemOffice is dependent on which property is studied, varying from good (pictures of the orbitals) to very good (renderings of the geometry), whereas Hyperchem and Spartan have overall a very good graphical display and output. ChemOffice and Spartan/Hyperchem differ significantly in other aspects: ChemOffice can take two-dimensionally drawn chemical structures (such as given in **Fig. 1**) and convert them into readable input structures for the computations, whereas the strength of Hyperchem and Spartan is that they include more advanced computational methods. Besides the semiempirical methods, they also include *ab initio* methods—such as Hartree-Fock (HF) or second order Møller-Plesset (MP2)—and density functional methods (DFT), including both relatively fast local DFT methods such as the Vosko-Wilk-Nusair method (VWN) and more accurate (and slower) gradient-corrected methods such as Becke-Perdew (BP) and its perturbational variation (pBP) (in Spartan).

The latter types of computations can also be performed with a wide variety of quantum chemical programs that are optimized for use with larger *ab initio* or DFT computations. The most well-known of these is Gaussian (http://www.gaussian.com), with which one can do almost all types of quantum chemical computations. This program does not provide a graphical user interface, but its output can be visualized easily with, for example, ChemOffice and especially Spartan. Other programs that can handle relatively large molecules such as flavins via *ab initio* or DFT methods include Jaguar (http://www.schrodinger.com) and Q-Chem (http://www.q-chem.com). The results of the latter can be visualized easily especially with Hyperchem.

2.2. Hardware

Molecular mechanics and semiempirical quantum chemical computations can conveniently be run on standard PCs (Pentium and later, with Windows 95/98/NT) or Macintoshes (PowerPC and later, with MacOS 7/8) using either ChemOffice, Hyperchem (PC only), or PC Spartan/MacSpartan. Minimum system requirements for efficient computations do usually include 16 MB of available physical RAM (besides what is used by the operating system). For *ab initio* or DFT computations relatively fast processors are required. Gaussian and Hyperchem are useful for this purpose on PII PCs, G3 Macintoshes, and later systems. This offers the advantage of the usage of the same operating system as used standardly on such machines (Windows 95/98/NT or MacOS 7/8). However, in the case of computations on molecules as large as flavins, UNIX machines are often preferable. Typical machines that can be used for this aim are Silicon Graphics computers such as the O2 or Octane, IBM/R6000 computers or DEC/Alpha. Such machines are the preferred type of hardware for use of Gaussian, Q-Chem, and Jaguar, and also for Spartan and Hyperchem if *ab initio* or DFT computations of systems of the size of flavins are frequently used options. The latter two programs can also run on fast PCs with adequate amounts of RAM (typically 64 MB minimal requirement, but 128 MB or more is advisable) and hard disk space (typically 2 GB minimal), given that Linux is used as the operating system. This is frequently a significantly cheaper alternative to the use of work stations, for all but the largest class of computations. At present no really stable version of Linux is well tested for the Macintosh. When Linux becomes available for Macintosh, all these programs can also be efficiently used on the G3 Power Macintosh machines, as their processor speed approaches that of standard UNIX-based work stations.

3. Method

The method for performing quantum mechanical calculations on flavins, but also on other molecules, includes the following basic steps:

1. Define the starting geometry.
2. Select a calculation model.
3. Define and select the output parameters required.
4. Run the calculation job.
5. Validate and analyze the results.

3.1. Define the Starting Geometry

The input geometries for calculations on flavin structures can be defined basically in two ways. First, the initial geometry can be designed by applying chemically reasonable atomic distances, angles and dihedral angles. In case of a text-oriented user interface (Gaussian) the distances, angles and dihedrals required to construct the molecule are entered in a so-called Z-matrix format. It is, however, often easier and faster to use a program with a graphical interface that provides a molecular editor, enabling the user to build the molecule interactively and including chemically reasonable bonds and angles automatically. These editors are sometimes supplemented with libraries of molecular fragments to further speed up the molecular building process. The structure built in this way can subsequently be submitted in the program, or saved in a format suitable for submission in another program, for example Gaussian.

The second approach to define the input geometry is to use existing structural data. For a growing number of compounds including proteins, crystal structure coordinates are available from the Cambridge crystallographic database (*dat* files) and the Brookhaven databank (*pdb* files), the latter containing the crystal structure coordinates of over 50 different flavin-containing proteins.

Such structure files or parts of these files, can be used as input by most programs, either by importing them directly, or after conversion to a recognizable format. Often, the crystallographic data do not contain hydrogen coordinates. These must be added by using either one of the building procedures mentioned above, or by using an automatic hydrogen building routine provided by a number of molecular modeling packages (e.g., ChemOffice, Insight II or Cerius^2).

3.2. Select a Calculation Model

Once the starting geometry has been defined one can use this structure for the quantum mechanical calculations. Selection of a calculation model involves considering the following questions.

1. Is the initial geometry relevant as such (e.g., in case of a crystal structure) or is (partial) geometry optimization required?
2. Is the system of interest charged? What is its multiplicity, i.e., how many unpaired electrons are present?
3. Is relevant information obtained from a calculation in vacuum or should the effect of the environment be taken into account?
4. Which quantum chemical method is best to use for the desired goal?

1. Is the initial geometry relevant as such (for example in case of a crystal structure) or is (partial) geometry optimization required? In many cases the input geometry may be relevant as such, for example when it is derived from an experiment (X-ray, nuclear magnetic resonance) or when it is optimized using another adequate computational method. A single point energy calculation is then sufficient to obtain all the electronic characteristics of the molecule.

Alternatively, when the initial structure is not based on available structural information, optimization of all atomic coordinates is preferable, leading to the energetically favorable geometry. Full geometry optimization is essential when the vibrational modes of the molecule are studied, for example in relation to spectroscopic properties, or in case of identifying a reaction coordinate.

In some cases geometric constraints can be helpful to obtain a relevant geometry. For example when the calculation is performed on an isolated fragment (e.g., an active site or a flavin molecule) on which, in reality, geometric constraints are imposed by the surrounding structure (e.g., an enzyme). For these purposes most programs provide possibilities to freeze atomic coordinates and to constrain atomic distances and angles to specific values during the geometry optimization. These constraint values can be derived on the basis of a relevant crystal structure or may be obtained from molecular mechanical calculations on the complete structure in which the isolated fragment is embedded.

2. Is the system of interest charged? What is its multiplicity, i.e., how many unpaired electrons are present? The number of electrons, which is an essential

input parameter for the calculation, is calculated by most programs on the basis of the nature and number of atoms in the molecule in combination with the charge defined. Another input parameter of interest, which is related to the number of electrons, is the number of unpaired electrons present. This characteristic is indicated by the multiplicity of the system (multiplicity = number of unpaired electrons + 1). This implies that when calculations on an anion, cation, or radical flavin species are of interest, the input for the calculations should define this by setting the correct charge and multiplicity values. By default most programs assume zero charge and the absence of unpaired electrons (multiplicity = 1). Charge and multiplicity numbers given should be compatible. Furthermore, the presence of unpaired electrons poses some additional requirements on the method to be used. By default most programs use the restricted Hartree-Fock method for singlet molecules, whereas unrestricted Hartree-Fock is used for molecules with one or more unpaired electrons.

3. Is relevant information obtained from a calculation in vacuum or should the effect of the environment be taken into account? Quantum chemical calculations are generally performed in vacuo (or "in gas-phase") i.e., assuming a dielectric constant of 1 and not including possible effects of the surroundings. Many properties of molecules can be derived from the in vacuo calculations, even if one is actually interested in their properties as solid, liquid or solute. Especially when a series of molecules is compared, in vacuo calculations often work well. However, in some cases it may be very important to include some effects of a specific environment. For example, polar solvents, with a high dielectric constant, reduce the electrostatic effects of local charges (dielectric screening) and drastically influence the energy of charged species. This effect is important when comparing differently charged molecules in solution or when energy differences are studied between intermediates of a reaction involving a redistribution of charge. A common way to take the effect of polar solvents into account is to include a so-called *reaction field* of uniform dielectric constant, surrounding the molecule. The polarization of this reaction field by the molecule results in an electric field, which in turn influences the electronic structure of the molecule *(1)*.

When more specific interactions with the surroundings are important, for example in the case of hydrogen bonds, it is preferable to include these interacting molecules explicitly in the structure to be calculated. However, the additional number of atoms, included in the quantum chemical calculation to model a specific environment, is limited due to rapidly increasing computational demands (as will be discussed in **Subheading 3.2., step 4**). In connection to this limitation it should be considered that when only a small part of, for example, an active site of an enzyme is included in the model, the geometric constraints imposed by the (rest of the) surrounding protein should be reproduced by well-chosen constraint parameters in the input (as discussed in **Subheading 3.2., step 1**).

A simplified approach to include the electrostatic effects of surrounding atoms, as for example implemented in the Gaussian program, is to define an approximate electrostatic potential field by means of fixed point charges around the molecule *(1)*. In more recently developed approaches, the environment surrounding the quantum chemical system, is fully modeled by a molecular mechanical treatment. In these so-called combined quantum mechanical and molecular mechanical (QM/MM) methods electrostatic as well as steric effects of the surrounding molecular model are included in the quantum chemical calculation (e.g., *2*). Although promising, these QM/MM methods are not yet widely used and various implementations are still in the process of development.

4. Which quantum chemical method is best to use for the desired goal? Quantum chemical methods vary in two aspects. First, different methods are present to handle the Hamiltonian operator in order to solve the Schrödinger equation. Second, the form and number of mathematical functions, which are used to describe the wave function and electron density of the molecule, are variable; these functions are organized in so-called basis sets.

Semiempirical methods, such as AM1 *(3)* and PM3 *(4)*, introduce a number of approximations in solving the Schrödinger equation to speed up the calculations drastically. Consequently, these methods can handle molecules containing up to 300 atoms relatively quickly, but results should be used qualitatively or to study relative effects within series of related compounds.

Ab initio methods offer the advantage of systematic convergence towards the exact solution of the Schrödinger equation via increase of electron correlation effects and basis set increase. However, this does not automatically happen in a straightforward way. Three methods can be commented on. Hartree-Fock computations are the most well-known of these, and give reasonable descriptions of both the geometry and electronic structure of many closed-shell molecules. For geometries, improvement over semiempirical methods for this type of species is usually small, so if geometrical information is the primary interest, use of semiempirical methods can save hours of CPU time. When chemical reactions, i.e., formation and breaking of bonds, are to be studied, the Hartree-Fock method is in principle not sufficient. Two improvements are possible in going from Hartree-Fock to methods with better accounts of electron correlation. Most classically this is done via perturbation theory, which is implemented in many programs via Møller-Plesset theory (specifically MP2). For neutral closed-shell species this gives a very good description of both geometry and electronic structures. The price is, however, a significant increase in computational demands, which renders this method usually completely impractical for systems of the size of flavins. A second, interesting approach consists of DFT methods. Especially the so-called hybrid-methods such as B3LYP are recommended for both geometrical and electronic structures, while also for quantitative agreement with experiment DFT methods often perform better than both semiempirical and *ab initio* HF methods.

Besides the applied method, the basis set used to describe the wave function is an important factor determining the quality of the calculation. The semiempirical methods all use a minimal basis set, containing the minimum number of basis functions needed for the valence electrons only. *Ab initio* methods can also be used with a minimal basis set, referred to as the STO-3G basis set, containing the minimum number of basis functions needed for all electrons. However, larger basis sets have been developed, imposing fewer restrictions on the spatial distribution of the electrons (and therefore resulting in a better approximation to the exact wave function). The first extensions to the minimal basis set are the so-called split valence basis sets, such as 3-21G, 6-31G, and 6-311G, which include two or more basis functions, with different sizes, for each valence orbital. In the 3-21G basis set, for example, each inner shell orbital is described by one basis function (obtained by combining 3 G functions), while the valence shell consists of two sets of basis functions (built from 2 and 1 G functions, respectively). These basis sets can be further extended by adding so-called polarization functions, i.e., basis functions of higher angular momentum quantum number than required to describe the ground state of each atom (e.g., p-type functions on hydrogens and d-type functions on C, O, and N, etc.) These polarization functions allow for the possibility of small displacements of electronic charge away from the nuclear positions and are denoted by adding the type of the additional function in brackets, e.g., 6-31G(d) or 6-31G(d,p), or with an asterisk, e.g., 6-31G* or 6-31G**, respectively. (A brief overview is given in *(1)*.)

Table 1 illustrates the performance of different methods and basis sets when applied on the flavin cofactor in its oxidized form with a CH_3 replacing the ribityl side chain. Full geometry optimization was performed, using a Silicon Graphics O^2 R10K workstation. **Table 1** also presents the basis functions, used to describe the H, C, O, or N atoms in the various basis sets.

From the data listed in **Table 1** it is clear that, with an increase in basis set, the time required for the calculation increases rapidly, thereby limiting the size of the molecules that can be handled by a specific method in the available amount of time. Clearly, one should choose a method by finding a balance between quality and costs. In the case of flavin molecules it is advisable, from a practical point of view, to start with the semiempirical models. They give a first, qualitative indication whether it is worthwhile to perform more time-consuming *ab initio* or density functional methods with more extended basis sets.

Table 1
Time Required for Full Geometry Optimization of the Oxidized Flavin by Different Quantum Mechanical Methods

Method	Basis set for H atoms	Basis set for C, O, N atoms	No. of basis functions	CPU time	Time relative to PM3
Semiempirical					
RHF/AM1	1*s*	2*s* $2p_x$ $2p_y$ $2p_z$	88	1 min 51 s	1.24
RHF/PM3	1*s*	2*s* $2p_x$ $2p_y$ $2p_z$	88	1 min 30 s	1.00
Ab initio					
RHF/STO-3G	1*s*	1*s* 2*s* $2p_x$ $2p_y$ $2p_z$	107	6 h 6 min	244
RHF/3-21G*	1*s* 1*s*	1*s* 2*s* $2p_x$ $2p_y$ $2p_z$ 2*s* $2p_x$ $2p_y$ $2p_z$	195	18 h 28 min	739
RHF/6-31G*	1*s* 1*s*	1*s* 2*s* $2p_x$ $2p_y$ $2p_z$ 2*s* $2p_x$ $2p_y$ $2p_z$ $3d_{xx}$ $3d_{yy}$ $3d_{zz}$ $3d_{xy}$ $3d_{xz}$ $3d_{yz}$	309	113 h 11 min	4527
Density functional					
pBP86/DN	1*s* 1*s*	1*s* 2*s* $2p_x$ $2p_y$ $2p_z$ 2*s* $2p_x$ $2p_y$ $2p_z$	195	23 h 47 min	951
pBP86/DN*	1*s* 1*s*	1*s* 2*s* $2p_x$ $2p_y$ $2p_z$ 2*s* $2p_x$ $2p_y$ $2p_z$ $3d_{zz}$ $3d_{zx}$ $3d_{yz}$ $3d_{xx}$ $3d_{xy}$	290	40 h 44 min	1629
pBP86/DN*	1*s* 1*s* $2p_x$ $2p_y$ $2p_z$	1s 2*s* $2p_x$ $2p_y$ $2p_z$ 2*s* $2p_x$ $2p_y$ $2p_z$ $3d_{zz}$ $3d_{zx}$ $3d_{yz}$ $3d_{xx}$ $3d_{xy}$	326	47 h 47 min	1911
BP86/DN*	as for pBP86/DN*		290	54 h 20 min	2173

[a]The input geometry was created using the build option in Spartan.
* = Polarization functions (*see* **Subheading 3.2., step 4**).

3.3. Define and Select the Output Parameters Required

Quantum chemical calculations may produce a large amount of output data. Therefore it is important to decide which output is relevant for the biochemical mechanism underlying the activity of the flavin cofactor studied. Many programs offer the possibility to select the data to be listed in the output and the analysis tools to be run at the end of the calculations. The selection proceeds either interactively or using keywords. It is advisable to choose not only the unknown characteristics of interest but also those electronic properties that allow comparison to existing (experimental) data.

Table 2
Quantum Chemical Parameters that can be Calculated, and Their Relation to Possible Characteristics of Flavin-Based Reaction Chemistry

Calculated parameter	Flavin-based reaction characteristic
Optimized geometry	Geometry of cofactor in its various reactive forms: flat, butterfly, or bent flavin ring, orientation of substituents
	Geometry of substrates
Energy of highest occupied molecular orbital [E(HOMO)]	Ionization potential of reduced or semiquinone flavin
	Nucleophilic reactivity of the cofactor or of substrates
Energy of lowest unoccupied molecular orbital [E(LUMO)]	Reduction potential of semiquinone or of oxidized flavin
	Electrophilic reactivity of the cofactor or substrates
Heat of formation	Differences in heat of formation between starting and transition state geometry or between starting and end geometry can be correlated to activation or reaction enthalpies, to the rate of a reaction, and/or to p*Ka* values
HOMO/HOMO-1 distribution	Indicates sites of nucleophilic reactivity within the cofactor and/or substrate
LUMO/LUMO+1 distribution	Indicates sites of electrophilic reactivity within the cofactor and/or substrate
Charge distribution	Relate to attractive/repulsive Coulomb interactions
Spin distribution	Relative stability of semiquinone flavin form, as spin delocalization stabilizes the radical
	Relates to electrophilic/nucleophilic reactivity of the sites within the radical
Bond orders	Indicates to what extent bonds are formed or broken, e.g., in reaction pathway geometries

Table 2 lists various of the parameters that can be calculated by quantum mechanical calculations, and links them to characteristics relevant for flavin-based reaction chemistry. In addition, some programs can generate graphical displays of the optimized geometries as well as of the calculated molecular orbital, spin and charge distributions. Although the numerical output is frequently most relevant, the visualization of abstract quantities is often helpful. In the **Notes** section several examples to illustrate the use of the various quantum mechanically calculated molecular orbital parameters for flavin-based research are presented.

3.4. Run the Calculation Job

After the starting geometry, method and all output to be generated have been defined the calculation can be started. Depending on the size of the molecule and the method chosen the calculation will take seconds, hours, days or even weeks. While running the calculation, the output generated reports the status of the calculation and can be monitored. When a geometry optimization is performed the output reports the results of each optimization cycle, for example, in terms of the energy gradient, which indicates to what extent the geometry is converging to an energy minimum. Since the geometry optimization is the most time-consuming part of many quantum mechanical calculations this also indicates when the calculation is about to be completed. When the convergence values fall beneath the optimization criteria defined in the program, the geometry optimization stops.

3.5. Validate and Analyze the Results

Validation and analysis of the results is an important step in the application of quantum mechanical calculations to biochemical problems. Since the programs started to be used by biochemists during the past decade, validation of the data obtained for biochemical systems is still an essential part of research in this field. Most of the effort is still devoted to comparing calculated geometries, reaction pathways, energy barriers and rates with data obtained from biochemical experiments. More and more calculations may then start to be of use not only as a tool for interpretation of experimental data, but also yielding additional information, for example when experimental data are unavailable because the short lifetime of reaction pathway intermediates. The next section includes some examples of validation of computational results in studies on flavin-dependent systems, showing the potential of this technique in flavin research.

4. Notes

1. Although some people tend to say that a computer experiment is always successful, this is not the case. First of all a calculation may fail due to inconsistencies in the input. For example, the multiplicity defined could be impossible for a given atomic structure and charge, the required parametrization or basis set may not be available for all the atoms of the system, or the method chosen may not be compatible with the size of the system, in relation to the computer memory and power available. Second, a calculation may fail when the geometry does not converge to an energy minimum. This may be related to the complexity of the molecule and the optimization routine implemented in the program. Some programs provide options or keywords to control some of the optimization parameters, such as for example the step-size and the maximum number of optimization steps required. When optimizations fail one might

consider the application of less precise optimization criteria, or use of a different input geometry.

Finally, the molecular orbital wave function may not converge to a "self-consistent" solution of the Schrödinger equation. This problem may be solved by defining a different initial guess wave function, obtained for example from a lower level calculation. In addition some programs provide a "damping" option that can be used to prevent oscillations in the self-consistent field procedure.

The computational failures discussed above are generally detected by the program and can often be solved. More serious failures, however, which require chemical knowledge to detect, relate to the fact that the generation of large output files does not necessarily imply the correctness of the data. Thus, a calculation can be "successful" but this does not guarantee that its results are "reasonable" and have a physical meaning. Several reasons implicit in the calculations themselves can be the cause of these failures. The following notes (**2–6**) discuss some problems or faults to be aware of, and strategies to solve and/or overcome these problems.

2. Geometry optimization: Geometry optimization is a critical aspect of the application of quantum chemistry on large systems. First, each optimization step may demand significant computational effort. Second, a large number of optimization steps may be required to find an energy minimum. Third, the optimized geometry could be a so-called local energy minimum, instead of a global energy minimum.

A common approach to limit the computation time for large molecules is to perform geometry optimization at a lower level of theory, followed by a single point calculation at a higher level of theory with a more extended basis set. Although this approach is often satisfactory, it should be noted that it does not always give good results. One should especially be aware of potential errors of the semiempirical methods in equilibrium geometries. Geometries representing an energy minimum on the semiempirical potential surface are possibly far from the corresponding minimum on the *ab initio* potential surface, as is illustrated in **Fig. 2**. Thus, results obtained from a single point calculation at the Hartree-Fock level using the geometry optimized with a semiempirical method may represent a distorted system and may not to be relevant for the equilibrium situation. On the other hand the optimum geometries obtained with *ab initio* methods, going from Hartree-Fock 3-21G* to higher level calculations, show a somewhat more systematic and often smaller variation. Therefore, higher level (e.g., MP2) energy calculations on Hartree-Fock 6-31G* geometries are generally believed to give reasonable results.

Another approach to limit the number of optimization steps is to perform partial optimization by fixing part of the system to its initial coordinates or by applying geometric constraints. However, on the basis of the arguments given above, the quality of the results depends largely on the initial geometry. Constraints on bond lengths and angles should in general be avoided in equilibrium geometry calculations, where CPU capacity allows this.

With increasing complexity of a molecule the chance to obtain a local minimum instead of a global energy minimum increases. Among the available

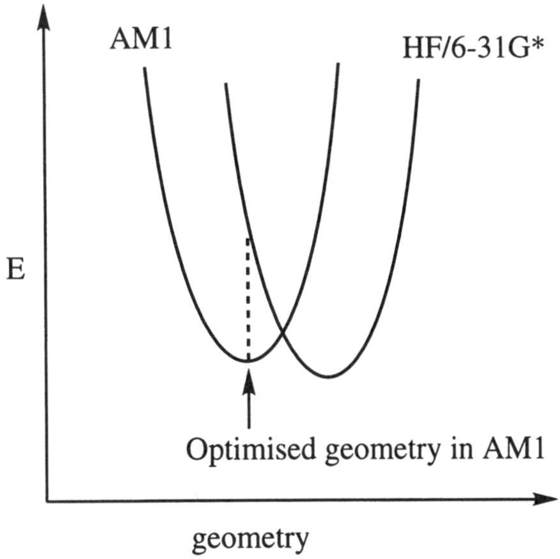

Fig. 2. The effect of different potential energy surfaces: an optimized geometry using a semiempirical method (e.g., AM1), may represent a distorted structure using the DFT or *ab initio* methods (e.g., HF/6-31G*).

methods to obtain a global energy minimum are Monte Carlo techniques and genetic algorithms that investigate and compare many possible conformations. In practice, all quantum mechanical methods, including even the semiempirical molecular orbital methods are likely to be too time consuming for these jobs and molecular mechanics methods should be applied. Then, the best geometries obtained by these molecular mechanical methods can be used as input for quantum mechanical energy minimizations. Alternatively, one may choose to use a relevant crystal structure geometry, since this geometry is (close to) a real life energy minimum. However, this cannot be done for transition state geometries for which crystal structural data are unavailable. Transition-state geometries need to be verified by means of a vibrational analysis, to check whether the vibration corresponding to the reaction coordinate (with an imaginary/negative eigenvalue) corresponds to the expected reaction. Even better, but often very time consuming, is to calculate an intrinsic reaction coordinate starting from the transition state, which should lead to correct reactant and product geometries.

3. Absolute vs relative outcomes: Computations on complex flavin systems can at present not be expected to yield reliable results for all calculated properties in an absolute way. Some calculated parameters, like bond lengths and angles, are frequently close to their physical values, but many electronic parameters are only useful on a relative basis. This implies that most calculations are to be used to study properties within a series of related structures, to correctly predict relative

differences. Experience obtained over the past decade has demonstrated that this relative approach, using series of related structures, is effective and can be used to study complex systems in a reliable way, even by the simple and fast semiempirical methods *(5–8)*. To illustrate this, the next paragraphs present some examples of quantum mechanical computer calculations in flavin biochemistry.

a. Redox potentials of a series of C7/C8-substituted flavins: The redox potential, i.e., the energy involved to donate or accept electrons, is an essential property of flavins since they are often involved in electron transfer reactions. The reduction potentials for a large number of flavin molecules have been reported in literature *(9–11)*. The energy required to take up an electron is to a first approximation related to the energy of the lowest unoccupied molecular orbital (LUMO). **Table 3** presents the experimental reduction potentials for the two-electron reduction of a series of C7/C8-substituted flavins as reported in the literature, as well as the LUMO energies calculated for these structures using various computational methods. Although the E_{LUMO} values obtained with the various methods differ significantly in an absolute way, they compare very well in a relative way (AM1 vs HF/3–21G*: $r = 0.96$; HF/3-21G* vs pBP/DN*: $r = 0.97$). Furthermore, with all three methods a clear correlation is observed between the calculated E_{LUMO} values and the experimentally determined reduction potentials of the uncharged oxidized flavin derivatives (**Fig. 3**). The results illustrate that when series of related compounds are studied, the absolute outcomes of the calculated parameters may deviate, whereas good quantitative structure activity relationships can still be obtained. The data also illustrate that the results from fast semiempirical methods can be very useful in these relative studies.

b. Calculated nucleophilic reactivity and activation barriers for a flavin-dependent monooxygenase reaction: According to frontier orbital theory, the energies of LUMO and HOMO (highest occupied molecular orbital) are not only related to reduction and oxidation potentials, respectively, but also to electrophilic and nucleophilic reactivity, i.e., the rate by which electrophilic and nucleophilic reactions proceed. As an example, the electrophilic attack by a C4a-(hydro)peroxyflavin cofactor on the C3 of the substrate parahydroxybenzoate, as catalyzed by the enzyme parahydroxybenzoate hydroxylase (PHBH), can be studied.

   First, the HOMO *density* on the individual atoms within the *p*-hydroxybenzoate flavin complex is related to their nucleophilic reactivity, relative to the other atoms. Comparison of the HOMO distributions within the substrate in two protonation states (**Fig. 4**) illustrates that deprotonation of the hydroxyl moiety is required to obtain substantial nucleophilic reactivity at the C3 position of *p*-hydroxybenzoate required for the electrophilic attack by the cofactor.

   Furthermore, the calculated HOMO *energies* of *p*-hydroxybenzoate and four of its fluorinated analogs correlate ($r = 0.99$) with the natural logarithm of their experimental rate of conversion *(13,14)*. Since the reactivity of the flavin cofactor is a constant factor in the enzymatic conversion of the series of

Table 3
Redox Potential for the Two-Electron Reduction of a Series of Oxidized Substituted Flavin Derivatives[a]

Substituent at		Reduction potential for two-electron reduction (mV)	E_{LUMO} (eV) calculated by		
C7	C8		AM1	HF/3-21G*	pBP86/DN*
CH_3	NH_2	–310/–330	–1.423	0.9951	–3.780
CH_3	OCH_3	–260	–1.700	0.8019	–3.938
CH_3	CH_3	–208/–207	–1.619	0.6993	–4.085
CH_3	H (nor)	–180	–1.647	0.6109	–4.199
CH_3	Cl	–152	–1.820	0.3720	–4.324
CH_3	F	–167	–1.821	0.4857	–4.264
CH_3	SCH_3	–204	–1.663	0.4234	–4.251
CH_3	$SOCH_3$	–161	–1.903	0.2416	–4.471
CH_3	SO_2CH_3	–50	–2.209	0.0607	–4.580
CH_3	H (iso)	–200	–1.620	0.6466	–4.131
CH_3	OCH_2CH_3	–246	–1.549	0.8245	–3.908
CH_3	Br	–148	–1.867	0.3959	–4.338
CH_3	$N(CH_3)_2$	–254	–1.569	0.7687	–3.810
H	H	–167	–1.683	0.5608	–4.286
H	CH_3	–199	–1.651	0.6607	–4.161
Cl	H	–131	–1.832	0.3094	–4.474
H	Cl	–145	–1.860	0.3167	–4.416
Cl	Cl	–98	–1.985	0.1097	–4.550
Cl	CH_3	–162	–1.793	0.4174	–4.332
F	H	–126	–1.878	0.2879	–4.474
H	F	–159	–1.861	0.4199	–4.376
F	CH_3	–158	–1.837	0.4076	–4.324
F	Cl	–104	–2.045	0.0680	–4.563
OCH_3	H	–156	–1.709	0.5442	–4.215
H	$N(CH_3)_2$	–340	–1.396	0.9791	–3.733
H	CN	–43	–2.118	–0.0797	–4.844
CN	H	–81	–1.960	0.1497	–4.748
CH_3	O^-	–334	2.225	4.6750	–0.199
CH_3	S^-	–290	1.596	3.7887	–0.542

[a]Obtained from the literature *(9–11)*, and the E_{LUMO} values calculated by several quantum mechanical calculations. For the calculations the N10-ethyl substituted derivatives were used. Reduction potentials reported by Hasford and Rizzo *(11)* were converted by +240 mV to be comparable to those reported by Eckstein et al. *(9)* and Ortiz-Maldonado et al. *(10)*.

substrates, the reactivity of the substrates systematically influences the rate of the reaction. Thus, calculations on the series of isolated substrates yield insight in the chemistry of the reaction process.

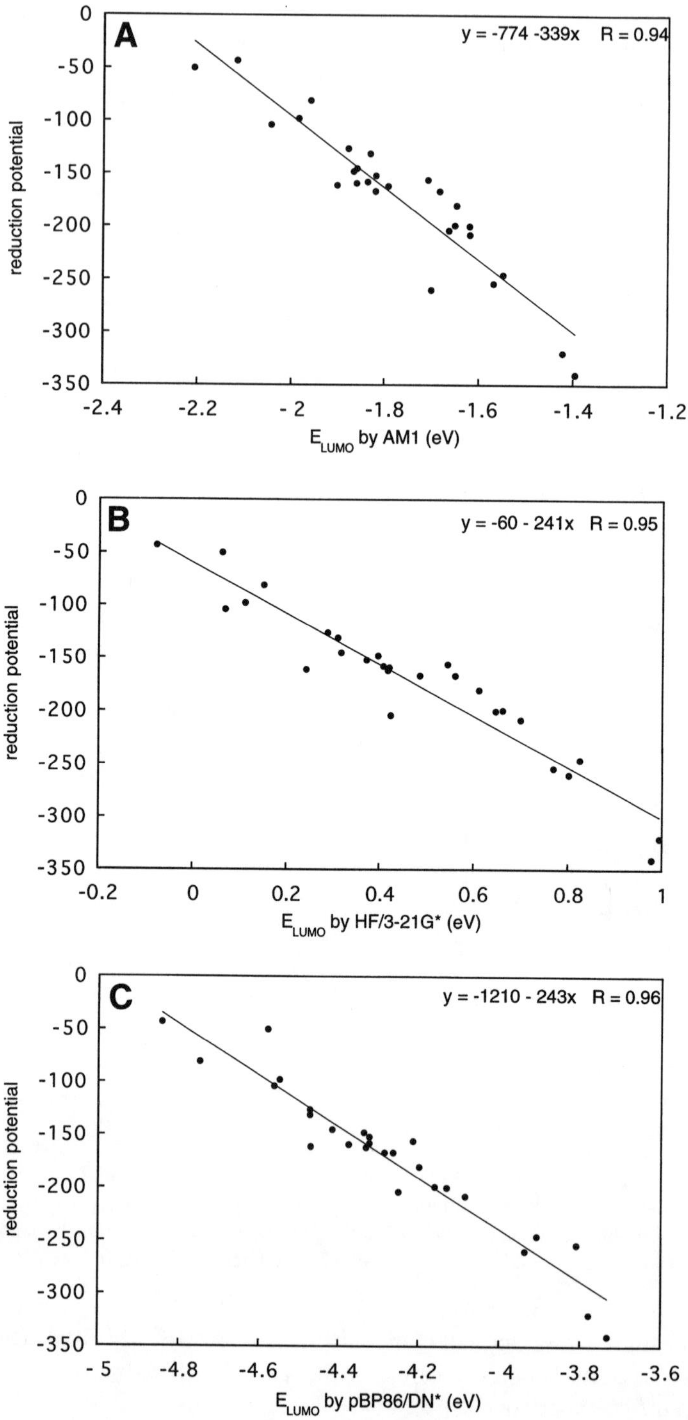
A
y = -774 -339x R = 0.94
reduction potential
E LUMO by AM1 (eV)
B
y = -60 - 241x R = 0.95
reduction potential
E LUMO by HF/3-21G* (eV)
C
y = -1210 - 243x R = 0.96
reduction potential
E LUMO by pBP86/DN* (eV)

Calculations are not restricted to structures and characteristics of stable molecules but can also be performed on reactive reaction pathway intermediates for which no experimental data for validation can be obtained. The lack of experimental data on transition state geometries complicates direct validation of transition state and reaction pathway calculations. As an example of reaction pathway calculations in flavin-based reaction chemistry we refer to recent QM/MM reaction pathway calculations for the monooxygenation of parahydroxybenzoate by the C4a-hydroperoxyflavin intermediate in the active site of para-hydroxybenzoate hydroxylase *(12)*. Such reaction pathway calculations do not yield absolute activation energies, since entropy is not included. However, the calculated energy barriers for the transfer of the OH moiety from the C4a-hydroperoxyflavin intermediate to the C3 of *p*-hydroxybenzoate and four of its fluorinated analogs, correlate well ($r = 0.96$) with the logarithm of the experimentally determined k_{cat} for conversion *(13)*, indicating they are useful in a relative approach. Since the fluorine substituent is relatively small, it has (almost) no steric effect, while it influences the electronic characteristics of a reactant to a large extent. Thus, such correlations with kinetic data allow validation of transition state and reaction pathway calculations.

It is important to note that the above-mentioned correlations are based on the assumption that the reaction of the C4a-hydroperoxyflavin cofactor with the substrate is the rate limiting step in the overall reaction. If not, another parameter than the calculated nucleophilicity of the substrate would determine the reaction rate.

4. Influence of charge: Comparison of properties in species with different overall charge requires caution. A striking example can be found in the comparison of the reduction potential data for two C8 substituted flavins, i.e., with $R = N(CH_3)_2$ (**Table 3**, entry 13) and $R = S^-$ (last entry). The E_{LUMO} of the S^- species is significantly higher than that of the species with $R = N(CH_3)_2$, yet the reduction potentials differ by only 50 mV.

   As previously explained, LUMO energy dominates the capacity of a compound to take up electrons. However, in case of a difference in overall charge two other terms come in for a correct comparison of the systems in solution. First, a difference in charge leads to a difference in solvation energy. The increase in solvation energy is usually different in going from an uncharged species to a negatively charged one, than for addition of a second electron to an already negatively charged species. Second, the addition of electrons has to overcome an increasing electron-electron repulsion, which is larger in going from a negatively charged species to a doubly negatively charged one, than in going from a neutral species to a charged one. These two factors usually cause the reduction of neutral

---

Fig. 3. Quantitative structure activity relationships for the correlation between the experimentally determined reduction potential for the two-electron reduction of a series of substituted flavins *(9–11)* and the E_{LUMO} as calculated by (A) the AM1 semiempirical method, (B) the HF/3-21G* method, and (C) the pPB86/DN* method.

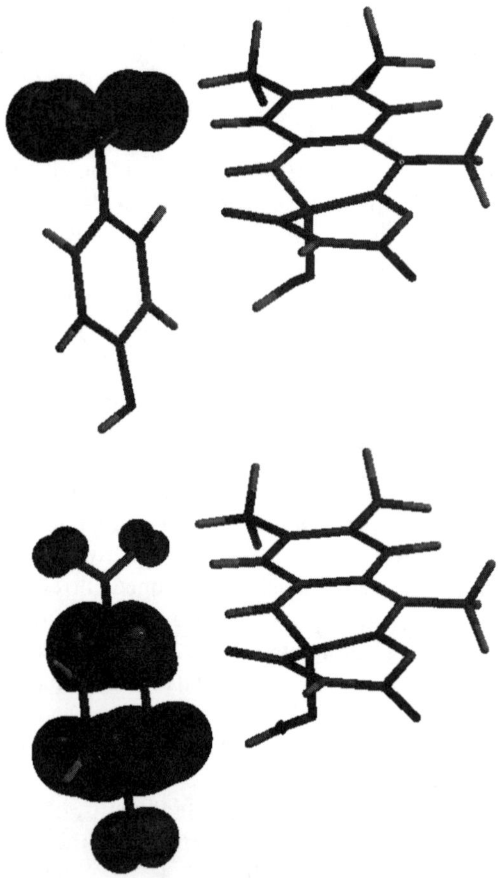

Fig. 4. Calculated (AM1) HOMO distribution in *p*-hydroxybenzoate flavin complex, with 1. a protonated OH moiety, and 2. a deprotonated O^- moiety. Note that upon deprotonation the HOMO density, representing nucleophilic reactivity, on C3 of the substrate parahydroxybenzoate increases from almost zero to 22%. The geometry is obtained from a combined QM/MM calculation *(12)*.

species to proceed more readily than that of analogous negatively charged species, and hamper the use of a single straightforward correlation for species with overall different charges.

5. Meaning of calculated charge distributions: The distribution of charge is not an unambiguously determinable molecular property. A variety of models are available, each with their specific advantages and disadvantages. The differences between several of these models can be very significant, so that differences in charge distributions between different species can only be meaningfully compared when using the same method. For the determination of how the charge of a species will affect its interaction with the surrounding medium (e.g., solvent or

protein), so-called electrostatic potential (ESP) charges are most significant, e.g., ChelpG or Merz-Kollmann (MK). For determination of the electronic structure of a species natural-orbital derived (NBO or NPA) charges are most reliable. Most programs do supply so-called Mulliken charges by default. These have as their only advantage that they are the most frequently used, but display notorious dependencies on theoretical methods and basis sets. Comparison of Mulliken charges obtained via different methods—even if only slightly different—renders them frequently completely useless. If ESP or NBO charges are available, these should be preferred for the specific aim chosen.

6. Influence of method used: It is common to assume that an increase of required computation time more or less automatically leads to an improved description of the system. This is, however, often not true in the comparison of the faster semiempirical methods to low-level *ab initio* calculations. In view of the size of flavin molecules, the applicability of the these fast methods on the problem of interest is to be considered first.

   The results of semiempirical methods are usually good in a qualitative sense, and very useful in the comparison of data obtained from structurally related compounds. However, in an absolute sense these methods tend to be less reliable for electronic properties, and also energetic differences between different conformations of a molecule should not be taken for granted using these methods. (In fact, if conformational analysis is of interest, molecular mechanics is usually the method of primary choice.) When choosing a specific semiempirical method one should take into account that the quality of the parameterization (AM1, PM3, etc.) varies for different elements. The elements C, H, N and O—as occurring in unfunctionalized flavins—are usually reasonably well described with the different semiempirical techniques. In contrast, for halogens, in which lone pair interactions can be substantial, the quality is generally improved in the series AM1 < PM3 < MNDO/d (vide infra). The latter treats C, H, N and O, however, via the less reliable MNDO method. Semiempirical methods are generally used within the framework of restricted Hartree-Fock calculations, in which orbitals are filled, starting from those with the lowest energy, with two paired electrons. For neutral molecules with closed electron shells (2 paired electrons in all orbitals) this is generally without problems, but for open-shell species (radicals, radical anions, radical cations or triplet states) this approach does have its limitations. In many programs the default treatment of open-shell species is via the unrestricted Hartree-Fock method, in which the functions describing α or β electrons are not necessarily identical anymore (although the resulting α and β function are usually still virtually identical). This approach works frequently reasonably well as long as the so-called spin contamination, i.e., S^2, is close to the theoretical value of 0.75 for a doublet spin system (i.e., a molecule with one unpaired electron), or 2.0 for a triplet system. Nearly all of the above mentioned programs will print values for S^2 by default in UHF computations. If the value is > 0.85 for a doublet system, the result should be seriously distrusted.

Semiempirical computations use so-called minimal basis sets. These have as a serious disadvantage that intermolecular interactions (e.g., π-π stacking interactions) are generally poorly described. Semiempirical computations compensate this limitation via element-specific parameterizations. In this way, PM3 approximates hydrogen-bonds to a frequently quite acceptable degree ***(15,16)***. For all species with elements containing lone pairs or (partially filled) d-orbitals, the results of semiempirical methods should be treated with caution, although some success has been claimed for methods including d-orbitals (MNDO/d) and for metals (PM3[tm]).

As indicated in **Subheading 3.2., step 4**, *ab initio* methods systematically converge towards the exact solution of the Schrödinger equation upon inclusion of electron correlation and increase of the size of the basis set. However, for large systems like flavin molecules, the size of the basis set should not be larger than necessary in relation to the problem studied. The use of *ab initio* and DFT computations with minimal basis sets (e.g., STO-3G) is strongly discouraged. The extra effort of *ab initio* or DFT calculations, compared to semiempirical calculations, only pays off with basis sets such as 3-21G(d), 6-31G(d) and larger. If hydrogen bonds play an important role, then polarization p-functions on hydrogen atoms are important, and increase to 6-31G(d,p) is recommended. Finally, for the description of negatively charged species, so-called diffuse functions—generally designated via the "+"-sign—should be included in the basis set. Convergence of the wave function with such basis sets is usually slowed down significantly, but electronic features are described more accurately. For anionic systems of the size of flavins the 3-21+G(d) basis would be recommended; if ample computational resources are available, then the 6-31+G(d,p) basis set is the next step. If one chooses to use diffuse functions in the description of anionic systems, direct comparison to neutral species requires computation of the neutrals with diffuse functions as well ***(17)***. For neutral species a single point calculation with diffuse functions at the optimized geometry obtained without diffuse functions is often adequate, as the inclusion of diffuse functions does not have major effects on the geometry of neutral species, but is important for electronic structure data and relative electronic energies of neutrals and anions. Further enlargement of the basis set via addition of more polarization functions is not paying off in most cases.

The exception here is formed by *ab initio* or DFT computations including (transition) metal atoms, which do demand large basis sets for accurate descriptions. Finally, the use of numerical basis sets is in some programs recommended for DFT work (e.g., DN in Spartan). Such basis sets might speed up the computation, but systematic validation of the quality of this approach, specifically for open-shell species, is currently not yet available.

5. Conclusions

The examples given in **Note 3** indicate that quantum chemical calculations become a useful tool in the field of flavin research. When possible, computational studies on flavins should be defined in a relative way, by investigation of trends in relevant electronic properties within a series of flavin molecules or with a series of substrates converted by the same flavin cofactor. In such problems, semiempirical methods are often very effective. Comparison of differently charged molecules is often problematic, especially in the solute state.

When computational results are to be reliable in an absolute sense, *ab initio* and density functional methods are preferable, provided that the basis set used and the electron correlation included give a sufficient description of the computational system. Furthermore, for absolute outcomes, one should consider the effect of the relevant environment on the properties of the molecule one is actually interested in. In biochemistry, the relevant environment is often the active site of a protein. A full description of all effects of a protein environment that surrounds a molecule of interest is beyond the current state of computational chemistry. However, the combined QM/MM methods as developed in the last decade are a first approach towards computational chemistry of biomolecules and, provided the results are used on a comparative basis, become an interesting tool in biochemistry.

References

1. Foresman, J. B. and Frisch, Æ. (1996) *Exploring Chemistry with Electronic Structure Methods,* Gaussian, Inc., Pittsburgh, PA.
2. Field, M. J., Bash, P. A., and Karplus, M. J. (1990) A Combined quantum mechanical and molecular mechanical potential for molecular dynamics simulations. *J. Comp. Chem.* **11,** 700–733.
3. Dewar, M. J. S., Zoebisch, E. G., Healy, E. F., and Stewart, J. J. P. (1985) AM1: a new general purpose quantum mechanical molecular model. *J. Am. Chem. Soc.* **107,** 3902–3909.
4. Stewart, J. J. P. (1989) Optimization of parameters for semiempirical methods. I. method. *J. Comp. Chem.* **10,** 209–220.
5. Cnubben, N. H. P., Peelen, S., Borst, J. W., Vervoort, J., Veeger, C., and Rietjens, I. M. C. M. (1994) Molecular orbital based quantitative structure-activity relationship for the cytochrome P450-catalyzed 4-hydroxylation of halogenated anilines. *Chem. Res. Toxicol.* **7,** 590–598.
6. Peelen, S., Rietjens, I. M. C. M., Boersma, M. G., and Vervoort, J. (1995) Conversion of phenol derivatives to hydroxylated products by phenol hydroxylase from *Trichosporon cutaneum;* a comparison of regioselectivity and rate of conversion with calculated molecular orbital substrate characteristics. *Eur. J. Biochem.* **227,** 284–291.
7. Van Haandel, M. J. H., Rietjens, I. M. C. M., Soffers, A. E. M. F., Veeger, C., Vervoort, J., Modi, S., Mondal, M. S., Patel, P. K., and Behere, D. V. (1996) Computer calculation-based quantitative structure-activity relationships for the oxidation of phenol derivatives by horseradish peroxidase compound II. *J. B. I. C.* **1,** 460–467.

8. Soffers, A. E. M. F., Ploemen, J. H. T. M., Moonen, M. J. H., Wobbes, T., Van Ommen, B., Vervoort, J., Van Bladeren, P. J., and Rietjens, I. M. C. M. (1996) Regioselectivity and quantitative structure-activity relationships for the conjugation of a series of fluoronitrobenzenes by purified glutathione *S*-transferase enzymes from rat and man, *Chem. Res. Toxicol.* **9,** 638–646.
9. Eckstein, J. W., Hastings, J. W., and Ghisla, S. (1993) Mechanism of bacterial bioluminescence: 4a,5-dihydroflavin analogs as models for luciferase hydroperoxide intermediates and the effect of substituents at the 8-position of flavin on luciferase kinetics. *Biochemistry* **32,** 404–411.
10. Ortiz Maldonado, M., Ballou, D. P., and Massey, V. (1997) Leaving group tendencies of 8-substituted flavin-C4a-alkoxides and the mechanism of hydroxylation catalyzed by p-hydroxybenzoate hydroxylase, in *Flavins and flavoproteins*, University of Calgary Press, Calgary, Alberta, Canada, pp. 323–326.
11. Hasford, J. J. and Rizzo, C. J. (1998) Linear free energy substituent effect on flavin redox chemistry. *J. Am. Chem. Soc.* **120,** 2251–2255.
12. Ridder, L., Mulholland, A. J., Vervoort, J., and Rietjens, I. M. C. M. (1998) Correlation of calculated activation energies with experimental rate constants for an enzyme catalyzed aromatic hydroxylation. *J. Am. Chem. Soc.* **120,** 7641–7642.
13. Husain, M., Entsch, B., Ballou, D. P., Massey, V., and Chapman, J. P. (1980) Fluoride elimination from substrates in hydroxylation reactions catalyzed by *p*-hydroxybenzoate hydroxylase *J. Biol. Chem.* **255,** 4189–4197.
14. Vervoort, J., Rietjens, I. M. C. M., Berkel, W. J. H. van, and Veeger, C. (1992) Frontier orbital study on the 4-hydroxybenzoate-3-hydroxylase dependent activity with benzoate derivatives. *Eur. J. Biochem.* **206,** 479–484.
15. Zuilhof, H., Lodder, G., and Koch, H. F. (1997) Carbon-oxygen hydrogen-bonding in dehydrohalogenation reactions: PM3 calculations on polyhalogenated phenylethane derivatives. *J. Org. Chem.* **62,** 7457–7463.
16. Jurema, M. W. and Shields, G. C. (1993) Ability of the PM3 quantum-mechanical method to model intermolecular hydrogen bonding between neutral molecules. *J. Comp. Chem.* **14,** 89–104.
17. Koch, H. F., Mishima, M., and Zuilhof, H. (1998) Proton transfer between carbon acids and methoxide: studies in methanol, the gas phase and *ab initio* MO calculations. *Ber. Bunsenges. Phys. Chem.* **102,** 567–572.

16

Flavins and Flavoenzymes in Diagnosis and Therapy

Katja Becker, Markus Schirmer, Stefan Kanzok, and R. Heiner Schirmer

1. Introduction

Riboflavin is a vitamin but not at all harmless. It is therefore not surprising that riboflavin and the riboflavinogenic coenzymes flavin mononucleotide (FMN) and flavin adenine dinucleotide (FAD) play a role in many fields of practical medicine. Here we report on new developments in medical flavinology focusing on the assessment of riboflavin status, medical interventions based on flavins, and current medical aspects concerning the enzymes glutathione reductase (GR) and thioredoxin reductase (TrxR) as representatives of a large family of homodimeric flavoproteins. For systematic and more specific information—particularly on the clinical aspects—we should like to refer the reader to comprehensive reviews ***(1–4)***. This article cannot deal with one specific method. However, methodological details of special medical interest are included. We will keep recalling a simple rule that is often broken in practical medicine and in medical research: *Drugs, B2 and light? The treatment can't be right*. It implies that flavins are readily photodegraded, and that a whole range of drugs ***(5,6)*** but also biological macromolecules ***(3,7,8)*** are subject to inactivating modification in riboflavin-dependent photoreactions. As exemplified for the cytostatic *Vinca* alkaloids vinblastine and vincristine as well as for the synthetic drug vindesine precautions have to be taken to prevent this side effect of riboflavin application ***(6)***.

The FAD-dependent enzymes glutathione reductase and thioredoxin reductase serve as examples for the implication of flavoproteins in diagnosis and therapy. Both enzymes represent important antioxidant principles in the cell and have been pointed out as promising targets for the design of cytostatic,

From: Methods in Molecular Biology, vol. 131: Flavoprotein Protocols
Edited by: S. K. Chapman and G. A. Reid © Humana Press Inc., Totowa, NJ

antiparasitic, and antirheumatic drugs *(9–11)*. GR and TrxR—obligatory homodimers—are particularly well suited for developing and testing compounds that interfere with protein folding and dimerization. As described here, glutathione reductase serves also as a model protein for studying the modifications which result from exposure to endogenous and pharmacologic nitric oxide (NO) donors *(12)*. Furthermore, in vitro saturation tests for erythrocytic flavoenzymes, in particular for glutathione reductase, are employed for the diagnosis of riboflavin deficiency *(1,13)*.

2. Materials

2.1. Determination of the Erythrocyte Glutathione Reductase Activation Coefficient (EGRAC)

1. Buffer A: 47 m*M* Potassium phosphate, 200 m*M* KCl, 1 m*M* EDTA, pH 6.9 at 25°C. 1 m*M* FAD in H_2O (final concentration 2 μ*M*) was used to activate the apoenzyme of glutathione reductase.
2. 4 m*M* NADPH and 20 m*M* glutathione disulfide were prepared in buffer A.

2.2. The Modification of Human Glutathione Reductase with NO Donors

1. *S*-Nitrosoglutathione was purchased from Alexis Biochemicals (San Diego, CA).
2. Dinitrosyl-iron-diglutathione complex (DNIC-$[GSH]_2$; 3.6 m*M*, stabilized with twentyfold excess of GSH) was obtained from Dr. Alexander Mülsch, Zentrum der Physiologie, Frankfurt University, Germany; it is kept at –70°C, and thawed only immediately before use.
3. Human recombinant glutathione reductase (GR) is prepared according to Nordhoff et al., 1993 *(14)*. The ε of the enzyme in oxidized form (E_{ox}) at 463 nm is 11.3 mM^{-1} cm^{-1} per subunit of 52.4 kDa.
4. GR is diluted, incubated, and measured in buffer A.
5. Enzyme activity is determined spectrophotometrically at 340 nm and 25°C in the presence of 0.1 m*M* NADPH and 1 m*M* GSSG in a total assay volume of 1 mL.
6. Solutions of GSSG (20 m*M*), NADPH (4 m*M*), and inhibitors in buffer A are prepared directly before use *(13)*.
7. For crystallizing GR modified with NO-donors, crystallization buffer (120 m*M* ammonium sulfate, 100 m*M* potassium phosphate, pH 7.5) is employed.

2.3. Reversible Denaturation of Glutathione Reductase and Thioredoxin Reductase from Humans and Malaria Parasites

1. Recombinant human glutathione reductase (190 U/mg) is produced and purified as described in **ref.** ***14***.
2. Authentic human thioredoxin reductase (39 U/mg in the DTNB assay) is isolated from placenta *(11)*.

Table 1
Renaturation Conditions Established for the Four Homodimeric Flavoenzymes Glutathione Reductase and Thioredoxin Reductase from Humans and the Malaria Parasite *Plasmodium falciparum*[a]

			Reactivation yield		
Enzyme	Protein	DTT	3 h	6 h	24 h
hGR	1 µg/mL	5 m*M*	58.3%	63.6%	68.2%
PfGR	2 µg/mL	5 m*M*	69.2%	70.5%	64.5%
hTrxR	10 µg/mL	1 m*M*	30.3%	31.7%	28.7%
PfTrxR	60 µg/mL	1 m*M*	51.4%	52.4%	57.7%

[a]After denaturation with 5 *M* guanidinium chloride, enzymes were diluted in 20 m*M* sodium phosphate, 0.5 m*M* EDTA, pH 6.9, to the respective protein concentration. Incubation of these samples in the presence of DTT resulted in partial reactivation of the enzymes; activities given in the table refer to controls that had not been exposed to 5 *M* GdnHCl. The EC_{50} of a folding inhibitor is the concentration needed to reduce reactivation yield at 3 h by a factor of 2.

Note that in the incubation mixtures for refolding and in the enzyme assay mixture the concentrations of residual GdnHCl and DTE are different for the four proteins. If other conditions or other enzymes are tested the influence of GdnHCl and DTE on the refolding protein and the assay must be reassessed.

3. *P. falciparum* thioredoxin reductase (4.8 U/mg in the DTNB assay) is produced and purified according to **ref. *15***. The clone was kindly provided by Dr. Sylke Müller, Bernhard-Nocht Institute of Tropical Medicine, Hamburg, Germany.
4. Recombinant *P. falciparum* glutathione reductase ***(10)*** had a specific activity of 160 U/mg. GR activity is determined in buffer A as described previously, TrxR activity is measured in buffer B (100 m*M* potassium phosphate, 2 m*M* EDTA, pH 7.4).
5. Denaturation buffer consists of 100 m*M* sodium phosphate buffer, 1 m*M* EDTA, pH 6.9, 2.5 m*M* DTT, 5 *M* ultrapure guanidinium chloride (GdnHCl).
7. Renaturation buffer consists of 0.1–5 m*M* DTT (*see* **Table 1**), 20 m*M* sodium phosphate buffer, 0.5 m*M* EDTA, pH 6.9, at 25°C.

3. Methods

3.1. Determination of the EGRAC as an Indicator of Riboflavin Deficiency

For the assessment of riboflavin status—which as a rule also reflects the general nutritional state of a person—the EGRAC represents a reliable quantity ***(1,16)***. Glutathione reductase (GR), an FAD-containing enzyme of 2 × 52.4 kDa, maintains a high cytosolic GSH/GSSG ratio by catalyzing the reaction GSSG + NADPH + $H^+ \rightleftharpoons$ 2 GSH + $NADP^+$. The FAD-free apoenzyme (apoGR) is stable *in situ* and can be complemented by FAD in vitro. The EGRAC is defined as FAD-stimulated GR activity divided by unstimulated GR activity which equals the ratio ([holoGR] + [apoGR])/[holoGR]. Based on the biochemical and structural properties of apoGR and holoGR, a highly sensitive and reliable version of the EGRAC test was developed; the test is adapted to the needs of tropical medicine and can be carried out with only 10 µL of finger-prick blood. Simultaneously, the hemoglobin concentration of the blood sample can be determined by absorbance measurements at 340 nm ***(13)***. The procedure is as follows:

1. 10 µL blood (venous or capillary) is mixed with 5 mL assay buffer A containing 0.04 m*M* digitonin as a cell lysing agent.
2. For duplicate determination of the EGRAC, 925 µL of this mixture is pipetted into each of 4 cuvets.
3. 2 µL 1 m*M* FAD is added to 2 of the cuvets in order to activate the apoenzyme of glutathione reductase.
4. After adding 25 µL 4 m*M* NADPH, the absorbance at 340 nm is monitored for 1 min to account for NADPH oxidase activity.
5. The GR-catalyzed reaction is then started in each cuvet with 50 µL 20 m*M* glutathione disulfide. The change in absorbance is monitored at 340 nm and 25°C over 10 min.
6. The EGRAC is then calculated by dividing ΔA/min of the samples with FAD (holoGR + activated apoGR) by ΔA/min of the samples without FAD (holoGR). Normal EGRAC values range from 1.0–1.4, an EGRAC higher than 1.4 is indicative of riboflavin deficiency ***(13,17)***. According to the equations

   $$\text{EGRAC} = ([\text{holoGR}]+[\text{apoGR}])/[\text{holoGR}] = 1 + [\text{apoGR}]/[\text{holoGR}] \text{ and}$$
   $$K_d = [\text{FAD}_f]\,[\text{apoGR}]/[\text{holoGR}] = 53\ \text{n}M$$

   the intraerythrocytic concentration of free FAD, the predominant riboflavinogenic coenzyme, can also be assessed ***(18)***:

   $$[\text{FAD}_f] = 53\ \text{n}M/(\text{EGRAC-1})$$

   It should be noted that an increase of the activation coefficient from 1.1 to 2.0 corresponds to a decrease of $[\text{FAD}_f]$ by a factor of 10.

When determining the EGRAC the following experimental details should be obeyed: The temperature of the cuvets should be preequilibrated to 25°C and kept constant during the time of the assay; due to the instability of the apoenzyme ($t_{1/2}$ = 25 min at 25°C) in dilute solutions, FAD must be added to the respective cuvets immediately after hemolysis. In blood samples where the absorbance at 340 nm decreases prior to GSSG addition, ΔA/Δt values must be corrected for this NADPH oxidase activity.

3.2. Interventions Based on Flavins and Flavoproteins

The methodological details of this section are too extensive to be described here; relevant references are given under **Subheading 4.2.** The following general precautions, however, apply to many clinical and experimental situations.

1. Flavin-containing solutions (for infusion or injection) and cell culture media must be protected from light by using dark bottles or by avoiding light < 500 nm.
2. Many compounds including drugs must be protected from riboflavin-triggered photoreactions. This applies, for instance, to doxorubicin- or vinblastine-containing solutions ***(5,6)***, and for drug solutions with riboflavin added as a compliance marker ***(19–21)***. Macromolecules like proteins, hyaluronic acid and DNA (in tissue cultures and *in situ*) should also not be exposed to flavin-mediated photoreactions ***(7,8)***. When considering a critical constellation of light, riboflavin, and the target molecule of an unwanted photoreaction it is often helpful to administer the flavin separately.

3.3. Modification of Human Glutathione Reductase by S-Nitrosoglutathione and Dinitrosyl-Iron-Diglutathione Complex

Signal transduction by nitric oxide (NO), the main component of endothelium derived relaxing factor (EDRF), plays a key role in biological systems; NO modifies, e.g., the bioregulation of vascular tone, platelet aggregation, neurotransmission, and macrophage activity (***22,23*** for literature reviews). The effects of the physiologic NO-donors *S*-nitrosoglutathione (GSNO) and dinitrosyl-iron-dithiolate complexes ***(12,24–26)*** on the flavoenzyme glutathione reductase have been analyzed in detail and serve as reference for the modification reactions with other proteins. Experimental conditions for enzyme inhibition and crystallization of the modified enzyme are given in the following paragraph.

1. To obtain inhibited enzyme for crystallization, recombinant hGR is exhaustively dialyzed against buffer A and then diluted with buffer A to 2 nmol GR subunits/mL. At higher enzyme concentrations the inhibition achievable with NO donors is less complete.
2. For reduction of the active site cystine, 1 m*M* NADPH is added.
3. Then the enzyme is incubated for 3 h with 1.5 m*M* GSNO or for 30 min with 36 μ*M* DNIC-$[GSH]_2$ until inhibition reached 85% and ≥ 95%, respectively. Owing to possible protein precipitation, the 85% GR-inhibition induced by GSNO should not be exceeded. Incubations for the inhibitor studies are carried out in closed plastic vials at 25°C in a water bath. Aliquots are taken at different time points for determining enzyme activity.
4. When the desired levels of inhibition are reached, 1 m*M* GSSG is added for oxidation of residual NADPH and the samples are dialyzed against crystallization buffer (*see* **Subheading 2.**).
5. Dialysis is followed by concentrating the inhibited enzyme to approx 10 mg/mL e.g., by Centriprep centrifugation (Amicon, Cleveland, OH) ***(12,24,25)***.

3.3.1. Crystallization of modified GR

In experiments prior to crystallization it had been shown that modification of hGR by the two inhibitors was possible in and not reversed by the presence of 2 *M* ammonium sulfate. Crystals were grown from the inhibited enzyme solutions using the following protocol. Hanging drops of 10 µL containing 13 mg/mL protein and 0.1% β-octyl glucoside in crystallization buffer (*see* **Subheading 2.**) were equilibrated with 700 µL reservoirs containing 720 m*M* ammonium sulfate and 100 m*M* potassium phosphate, pH 8.0. For storage and handling, crystals were transferred to a stabilizing solution containing 1 *M* ammonium sulfate and 100 m*M* potassium phosphate, pH 8.0 ***(12)***.

3.4. Denaturation and Reactivation of Glutathione Reductase and Thioredoxin Reductase of P. falciparum and Humans

Human glutathione reductase is an FAD-containing antioxidant enzyme of known geometry ***(27)***. According to structural predictions based on deduced amino acid sequences, three other proteins are very likely to have a three-dimensional structure similar to that of human GR: Human thioredoxin reductase, *P. falciparum* glutathione reductase (PfGR) and *Plasmodium falciparum* thioredoxin reductase (PfTrxR) ***(10,15,28,29)***. All four enzymes represent potential targets of antimalarial and cytostatic agents, respectively. Since these four proteins are obligatory homodimers, dimerization inhibition represents a promising approach to drug design. In order to test such inhibitors, assays based on unfolding and subsequent refolding of the 4 enzymes have been established:

1. GR-activities are measured spectrophotometrically (340 nm, 25°C) in buffer A containing 0.1 m*M* NADPH and 1 m*M* glutathione disulfide. TrxR-activity is determined at 25°C in buffer B containing 0.2 m*M* NADPH and 3 m*M* DTNB as a disulfide substrate. The relevant ε-value for the product, 2-nitro-5-thiobenzoate, is $2 \times 13.6\ mM^{-1} \times cm^{-1}$ at 412 nm.
2. Denaturation of all 4 proteins is carried out in denaturation buffer (*see* **Subheading 2.**). Proteins at 0.3–1.0 mg/mL are incubated for 15 min at 20°C in closed plastic vials. Denaturation is measured by loss of enzyme activity that cannot be reversed by dilution within minutes. For hGR it has been shown previously by CD spectroscopy that loss of activity after incubation with GdnHCl correlates with denaturation of the protein.
3. Renaturation of the enzymes is achieved by dilution of the denatured samples in renaturation buffer with DTT concentrations varying from 0.1–5 m*M*. Stirring was found to be critical. After denaturation the samples are stirred magnetically with an excess of renaturation buffer for exactly 10 s. Enzyme (range 1–60 µg/mL) and DTT concentrations (0.1–5 m*M*) as well as incubation times (6 min to 24 h) were varied systematically in order to find the optimal conditions for renaturation.

4. The success of renaturation is measured by determining enzyme activity. **Table 1** summarizes the optimal renaturation conditions *(30)*. As shown for hGR, FAD in the renaturation buffer (1 μ*M* to 100 μ*M*) or in the enzyme assay mixture (5 μ*M*) does not increase speed or yield of reactivation. On the contrary, the highest activity was measured when the renaturing sample exclusively contained the FAD which had been released during unfolding *(31)*. Thus, apoglutathione reductase, the folded FAD-free protein, is unlikely to be an intermediate of the renaturation pathway. Rather, FAD is probably incorporated early in the folding process.

4. Notes

4.1. The EGRAC as an Indicator of Riboflavin Status

4.1.1. Field Studies

The EGRAC has been employed in a large number of clinical studies; two of them shall be given as examples: A study of Bates et al. (1994) showed that in rural Gambia the riboflavin status as indicated by the EGRAC strongly depends on the season *(17)*. The activation coefficient is highest (close to 2.0) from December to March and down to < 1.5 in July. These data are likely to be explained by a changing dietary riboflavin intake. As shown for pregnant women in Northeast Thailand, the EGRAC does not change during the course of pregnancy; in contrast, a significant EGRAC difference between women living in urban (1.04–1.09) and rural areas (1.23–1.28) was detected *(32)*.

4.1.2. False Negative and False Positive Results

Up to now, three (patho)physiological conditions have been described which can—apart from riboflavin status—influence the EGRAC-value. As described by Flatz *(33)* who carried out a study in Thailand, subjects with glucose-6-phosphate dehydrogenase (G6PDH) deficiency exhibited significantly lower EGRAC-values (mean = 1.13) than control persons without this enzyme defect (mean = 1.79). There were no indications for a better nutritional status of the first group. These data suggest that the dissociation constant of holoGR is decreased in G6PDH-deficiency. It remains to be studied whether in vivo interactions between GR and G6PDH are responsible for conformational changes of the flavoenzyme leading to facilitated FAD release. For assessing the riboflavin status in G6PDH deficient persons—more than 80 million families are affected worldwide—another flavoenzyme complementation test is recommended *(1,3)*.

A second condition under which misleading EGRAC-values are observed is, unexpectedly, represented by severe protein-energy malnutrition *(34)*. In a clinical study carried out in Nigeria, EGRAC-values of children suffering from the malnutrition syndrome kwashiorkor, which is characterized by generalized edema, skin lesions, and infections, were compared to those of marasmic, severe marasmic, and control children. In the control group, a mean EGRAC of 1.52 was measured indicating a moderately insufficient riboflavin intake of Nigerian children. In marasmic children the EGRAC (1.39) was slightly elevated but within the normal range; in children with severe marasmus and kwashiorkor, "normal" EGRAC-values (1.22 and 1.25) were detected. However, these results are unlikely to reflect an adequate riboflavin intake, they are rather indicative of muscular wasting taking place in severe malnutrition: riboflavin released from muscle tissue can normalize the FAD status of other cells including erythrocytes ***(34,35)***.

The in vivo conversion of riboflavin to FMN and FAD is controlled by thyroid hormones ***(36,37)***. A study carried out with blood samples from 33 well-nourished thyroidectomized patients and 21 euthyroid control persons demonstrated significantly higher EGRAC-values in the patients (1.40 vs 1.22) indicating that the concentration of free FAD is approx 2 times lower in the erythrocytes of hypothyroid persons than in the controls ***(13)***. Therefore, the EGRAC can also be considered as a metabolic parameter in the diagnosis of hypothyroidism; on the other hand, false positive results can be obtained in hypothyroid patients if the EGRAC is used as an index of riboflavin deficiency.

4.1.3. Riboflavin Oversupplementation in Premature Infants

Despite the potential dangers of over- and undersupplementation with riboflavin, little is known about status, requirements and metabolism of riboflavin in very low birthweight infants. We therefore determined the EGRAC in a cohort of preterm babies (median of birthweight 634 g, median of gestational age 24 wk +3 d) at intervals over a period of 6 wk ***(38)***. All infants were in an intensive care unit and received riboflavin supplementation. The average EGRAC was 1.06 which corresponds to [FAD_f] of approx 900 n*M*; this value is 4–5 times higher than in cord blood or in breast-fed infants. The accumulation of FAD is paralleled by highly increased plasma riboflavin concentrations ***(39)*** and can be explained by a limited renal capacity of preterm infants to eliminate excess riboflavin. Established risks of B_2-hypervitaminosis range from kidney toxicity to light-mediated formation of reactive oxygen species with subsequent damage to DNA and other macromolecules. The latter effect is, of course, enhanced by phototherapy ***(40)***. Our study ***(38)*** has led to dose reduction for B_2-supplementation in preterm infants.

In conclusion, the activation coefficient of erythrocyte glutathione reductase is a reliable parameter of riboflavin status. The EGRAC has to be cautiously interpreted in persons with G6PDH-deficiency or hypothyroidism, and in extremely malnourished children ***(1,13,33,34)***. It might, however, also be a useful parameter for course control of conditions which alter riboflavin or GR metabolism—like substitution therapy in hypothyroidism, phototherapy, the use of certain contraceptives, and the administration of the cytostatic agent carmustine or tricyclic antidepressant drugs ***(3)***. Carmustine leads to irreversible modification of holoGR and holoTrxR in vivo. As shown for GR, the apoenzyme is not significantly attacked in vivo. The desired EGRAC value during chemotherapy is above 4. Isolated apoGR in vitro, however, is easily inhibited by nitrosureas ***(41)***.

4.2. Interventions Based on Flavins and Flavoproteins

4.2.1 Correcting Riboflavin Deficiency

The recommended daily intake of riboflavin for adults is about 2.0–2.5 mg ***(1)***. This can be exceeded in riboflavin-responsive inherited diseases ***(3,42)*** and conditions like myocardial reperfusion injury ***(43)***. Physiologically, excess riboflavin is excreted by the kidney. However, as noted previously, in premature infants and patients with renal failure flavin oversupplementation must be avoided.

Flavins (riboflavin, FMN and FAD) are of particular importance in reproductive medicine. Fatty acid catabolism and thermogenesis of the newborn are based on an ensemble of flavoenzymes ***(1,3)***. Moreover, placenta formation probably depends on a specific riboflavin carrier protein (RCP) which must be saturated with riboflavin ***(44)***. This means that women of child-bearing age should have an adaequate supply of vitamin B_2. In this context it is worth mentioning that reproduction-specific RCP is also studied as a potential antifertility vaccine in mammals and possibly in humans ***(45)***.

Essential flavoenzymes—like cytochrome P450 reductases, respiratory burst NADPH oxidase and inducible NO-synthases—are involved in detoxification of chemicals and protection from infectious agents; these enzymes do have standby functions and can be up-regulated when needed. In addition, cytochrome P450 reductase and iNOS, have not only FAD but also FMN as a prosthetic group. This should be accounted for, since the levels of FMN are affected more drastically in riboflavin deficiency than those of FAD. Persons exposed to specific hazards, for instance, workers in chemical plants with inadequate safety standards, must therefore have a guaranteed riboflavin intake ***(3)***.

4.2.2. Riboflavin-Responsive Inherited Diseases

There is a growing list of pathological conditions where high-dosed riboflavin administration shows a beneficial or even curative effect ***(1–3)***. In most cases the biochemical basis of the disorder and of the therapeutic success appears to be a pathologically high dissociation constant of an individual FAD- or FMN-dependent protein. If the level of a given protein is pathologically low, high levels of the flavin cofactor will also, according to mass action, increase the proportion of active protein. Detailed studies on patients with muscular weakness caused by riboflavin-responsive complex I deficiency have been published in recent years ***(42,46)***. Other cases—including patients suffering from epilepsy and other neuromuscular disorders, habitual early abortion, disorders of iron metabolism, and lens cataract could also be traced back to low flavin saturation of proteins such as fatty acylCoA dehydrogenases ***(47)***, reproduction-specific riboflavin carrier protein, NADPH-dependent flavin reductase, and glutathione reductase ***(3)***. These analyses have led and will lead to new insights into the interaction of apoflavoproteins and their cofactors and thus to new therapeutic strategies.

In systematic studies with larger groups of patients, beneficial effects of high-dosed riboflavin given for approximately one year were observed for patients with migraine ***(48)*** and in tobacco chewers who have a high risk of developing precancerous lesions in the upper aerodigestive tract (***49***; *see* also ***50***).

4.2.3. Use of High-Dose Riboflavin as a Marker for Monitoring Medication Compliance of Patients

Adherence to prescribed medication is a major public health problem. As shown in recent studies, riboflavin administered as a drug additive represents a reliable objective measure for medication-taking behavior. The underlying biochemical principle is as follows. Riboflavin in excess of the daily requirement of 2.5 mg is excreted by the kidney and can be detected in urine by using a fluorescence lamp. In practice, 50 mg riboflavin administered at night can be readily identified in urine over the next 24 h ***(19–21)***. It should be noted that riboflavin is used here for a diagnostic procedure and it should be stressed again that the mixture of drugs and riboflavin must not be exposed to light below 500 nm before use. Otherwise the following constellation is possible: The physician knows that the patient has taken his drugs but these drugs had been chemically modified by riboflavin-triggered photoreactions!

4.2.4. Photochemistry of FMN in Microsurgery

Microsurgery offers another photochemical aspect of flavins: In eye and ear surgery it is often compulsory to “solder” wounds during the operation. For this purpose Khadem et al. ***(51)*** have tested a number of heat- and light-activated tissue glues. Using the “large cornea wound” as a model, the authors favor the mixture of fibrinogen and FMN as a photodynamic biological tissue glue. When this glue is activated by a blue-green argon laser the flavin serves as a generator of reactive oxygen species which cross-link the fibrinogen with corneal stroma collagen. When challenged immediately after soldering, the wounds were found to resist mechanical strains exceeding the arterial blood pressure.

4.3. Effects of Physiological NO Donors at the Catalytic Site of Glutathione Reductase

Nitric oxide (NO) plays an important role in many (patho)physiological processes and is likely to exert some of its effects via modulation of enzymatic activities. The growing list of enzymes modified by NO(carriers) includes guanylate cyclase, glyceraldehyde-phosphate-dehydrogenase, ribonucleotide reductase, alcohol dehydrogenase, glutathione transferase, and cytochrome c oxidase ***(23)***. Effects of the physiological NO-carrier molecules *S*-nitrosoglutathione ***(12,24)*** and dinitrosyl-iron-diglutathione complex (DNIC-$[GSH]_2$) ***(12,25,26)*** on the flavoenzyme glutathione reductase have also been described; only for this enzyme the resulting modifications are known in atomic detail.

After 2 h incubation with 1 m*M* GSNO, crystalline human erythrocyte GR (2.5 U/mL, representing intraerythrocytic concentrations) is inhibited by approx 85%. The inactivating modification depends on the presence of NADPH and can only be reversed by incubation with high concentrations of thiols, for instance with 100 m*M* DTE at pH 7.5 for 2 h. GSNO is no substrate in the forward reaction of GR (which implies that *S*-nitrosoglutathione cannot be accounted for by methods that employ GR for determining total glutathione). When tested as a reversible inhibitor under assay conditions GSNO competes with GSSG the K_i being 0.5 m*M* ***(24)***.

Human GR was also found to be inhibited by dinitrosyl-iron-diglutathione (DNIC-$[GSH]_2$). The inhibition takes place in two ways: (1) the inhibitor is *competitive* with GSSG the K_i being approx 5 μ*M*; (2) when preincubating GR for 10 min with the compound, 4 μ*M* DNIC-$[GSH]_2$ leads to 50% *irreversible* enzyme inhibition. More than 95% GR-inactivation was achieved by incubation with 36 μ*M* for 30 min. This inhibition also depends on the presence of NADPH ***(25,52)***.

Spectroscopic studies of the modified proteins are based on the fact that unmodified glutathione reductase exists in two stable states: E_{ox} is a yellow-colored species with an active site disulfide (Cys58-Cys63), and EH_2 is the NADPH-reduced orange-red dithiol form of the enzyme. This color change is due to a charge transfer complex between the active site thiolate Cys63 and flavin. Absorption spectra of GSNO-modified GR as well as of DNIC-$[GSH]_2$-inhibited GR showed that the charge-transfer interaction between the isoalloxazine moiety of the prosthetic group FAD and the active site thiol Cys63 is disturbed by the modification; in GSNO-modified GR—but not in DNIC-modified GR—the charge transfer band around 530 nm can be regained by incubation with dithiothreitol (> 20 m*M*) for several hours. Indeed, as demonstrated by X-ray crystallography, Cys63 is oxidized to a sulfenate in GSNO-modified GR and to a sulfinate in DNIC-$[GSH]_2$-modified GR. Cys58 is present as a mixed disulfide with glutathione in both inactivated enzyme species *(12)*.

Microinjected GR crystals have been used as redox indicators of subcellular compartments *(52)*. It will be of interest to test whether or not these crystals can also be employed for studying nitrosative stress in vivo. We predict that the color of injected microcrystals will change from yellow to red in unchallenged cells whereas yellow crystals with modified Cys63 will result when the enzyme becomes exposed to NO and NO carriers.

4.4. Denaturation/Reactivation-Based Assays for Folding Inhibitors of GR and TrxR from P. falciparum and Humans

Glutathione reductase (GR; EC 1.6.4.2) has been studied extensively as a target of cytostatic and antimalarial compounds *(9,53)*. Catalyzing the reaction NADPH + GSSG + $H^+ \rightarrow NADP^+$ + 2 GSH, the homodimeric GR is the central enzyme of the glutathione redox metabolism. The best known GR inhibitor—which also inactivates thioredoxin reductase *(54)*—is carmustine (BCNU), a widely used antitumour drug *(41)*. Of special interest to the flavinologist are riboflavin analogs as precursors of FAD analogs *(55,56)* and 10-aryl-iso-alloxazines. The latter compounds—most promising antimalarial agents *(57,58)*—do not bind at the FAD site but in a cavity between the two subunits of glutathione reductase *(58,59)*.

Here we report on a test for inhibitors that interfere with enzyme folding or dimerization. In order to test this new type of inhibitors, an assay for denaturation and refolding of human GR was established by Nordhoff et al. *(31)*. These studies have now been extended to three other GR-like proteins which represent interesting potential drug targets: Human thioredoxin reductase (hTrxR) *(11)*, *P. falciparum* glutathione reductase (PfGR) *(10,28)*, and a putative *P. falciparum* thioredoxin reductase (PfTrxR) *(15)*. For each enzyme denaturation and reactivation conditions have been systematically optimized; this was necessary because successful folding studies on oligomeric enzymes from eukaryotes are rare; it is quite possible that the folding pathway for these proteins is more complex than for oligomeric proteins of viruses and prokaryotes. Most studies on folding inhibition were done on human GR. Since both subunits contribute essential residues to both active sites the monomer of the enzyme is inactive. Consequently it is possible to prevent formation of active enzyme by inhibiting dimer formation. *Ab initio* folding and/or dimerization of hGR can be inhibited by mutations *(14,30)* or by peptides corresponding to subunit-interface areas. A point in case are synthetic peptides which represent the intertwining contact helices H11 (residues 439–453) and H11' (residues 439'–453') of human GR *(31)*.

P11, a peptide corresponding to residues 436 to 459 of GR, inhibited human GR dimerization with an $EC_{50\%}$ of 20 μ*M* ***(31)***. Among the shorter peptides of this region, one was found to be effective in submicromolar concentrations ***(30)***. The solution structure of the synthetic peptide was solved by ^{1}H NMR spectroscopy. Surprisingly, the peptide showed the same well-ordered helical structure in solution as in the intact protein ***(27)***; the two structures are almost perfectly superimposable ***(31)***.

With P11 and segments of this peptide as lead compounds, it might be possible to develop peptidomimetic inhibitors of proteins such as hGR, PfGR, hTrxR, PfTrxR, and other related enzymes like lipoamide dehydrogenases from different species, trypanosomal trypanothione reductase, and bacterial NADH oxidases ***(60)***. Due to their high specificity and possible effects in substoichiometric amounts, dimerization inhibitors are promising candidates for the design of antiparasitic, cytostatic and antirheumatic drugs. Their detailed mechanism of action can be followed by studying their effects on fluorescence and absorption of FAD during protein folding.

Acknowledgments

The authors wish to thank Stephan Gromer for helpful discussions. Our work is supported by the Deutsche Forschungsgemeinschaft (Be 1540/4-1) and by the Fouds der Chemischeu Industrie, Gaut Nucuber 151676.

References

1. Bates, C. J. (1987) Human riboflavin requirements and metabolic consequences of deficiency in man and animals. *World Rev. Nutr. Diet.* **50,** 215–265.
2. Müller, F., Ghisla, S., and Bacher, A. (1988) Vitamin B_2 und natürliche Flavine, in *Vitamine II* (Isler, O., Brubacher, G., Ghisla, S., and Kraeutler, B., eds.), Georg Thieme Verlag, Stuttgart, Germany, pp. 50–159.
3. Krauth-Siegel, R. L. and Schirmer, R. H. (1991) Flavoproteins in Medicine, in *Chemistry and Biochemistry of Flavoenzymes*, vol. I (Müller, F., ed.), CRC Press, Boca Raton, FL, pp. 275–286.
4. Becker, K., Keese, M., Gromer S., and Schirmer, R. H. (1997) Flavins in medicine (minireview). *Flavins Flavoproteins* **12,** 3–12.
5. Bomgaars, L., Gunawardena, S., Kelley, S. E., and Ramu, A. (1997) The inactivation of doxorubicin by long ultraviolet light. *Cancer Chemother. Pharmacol.* **40,** 506–512.
6. Granzow, C., Kopun, M., and Kröber, T. (1995) Riboflavin-mediated photosensitization of *Vinca* alkaloids distorts drug sensitivity assays. *Cancer Res.* **55,** 4837–4843.
7. Frati, E., Khatib, A. M., Front, P., Panasyuk, A., Aprile, F., and Mitrovic, D. R. (1997) Degradation of hyaluronic acid by photosensitized riboflavin *in vitro*. Modulation of the effect by transition metals, radical quenchers, and metal chelators. *Free Rad. Biol. Med.* **22,** 1139–1144.

8. Mori, T., Tano, K., Takimoto, K., and Utsumi, H. (1998) Formation of 8-hydroxyguanine and 2,6-diamino-4-hydroxy-5-formamidopyrimidine in DNA by riboflavin mediated photosensitization. *Biochem. Biophys. Res. Commun.* **242,** 98–101.
9. Schirmer R. H., Müller J. G., and Krauth-Siegel R. L. (1995) Disulfide-reductase inhibitors as chemotherapeutic agents: The design of drugs for trypanosomiasis and malaria. *Angewandte Chemie Int. Ed. Engl.* **34,** 141–154.
10. Färber, P. M., Arscott, L. D., Williams, C. H., Becker, K., and Schirmer, R. H. (1998) Recombinant *Plasmodium falciparum* glutathione reductase is inhibited by the antimalarial dye methylene blue. *FEBS Lett.* **422,** 311–314.
11. Gromer, S., Arscott, L. D., Williams, C. H., Schirmer, R. H., and Becker, K. (1998) Human placenta thioredoxin reductase. Isolation of the selenoenzyme, steady state kinetics, and inhibition by therapeutic gold compounds. *J. Biol. Chem.* **273,** 20,096–20,101.
12. Becker, K., Savvides, S., Keese, M., Schirmer, R. H., and Karplus, P. A. (1998) Enzyme inactivation by NO carrier-based sulfhydryl oxidation. *Nature Struct. Biol.* **5,** 267–271.
13. Becker, K., Krebs, B., and Schirmer, R. H. (1991) Protein-chemical standardization of the erythrocyte glutathione reductase activation test (EGRAC test). Application to hypothyroidism. *Int. J. Vitam. Nutr. Res.* **61,** 180–187.
14. Nordhoff, A., Bücheler, U. S., Werner, D., and Schirmer, R. H. (1993) Folding of the four domains and dimerization is impaired by the Gly446' → Glu exchange in human GR. Implications for the design of antiparasitic drugs. *Biochemistry* **32,** 4060–4066.
15. Müller, S., Gilberger, T. W., Färber, P. M., Becker, K., Schirmer, R. H., and Walter, R. D. (1996) Recombinant putative glutathione reductase from *Plasmodium falciparum* exhibits thioredoxin reductase activity. *Mol. Biochem. Parasitol.* **80,** 215–219.
16. Glatzle, D., Weber, F., and Wiss, O. (1968) Enzymatic test for the detection of a riboflavin deficiency. NADPH-dependent glutathione reductase of red blood cells and its activation by FAD in vitro. *Experientia* **24,** 1122.
17. Bates, C. J., Prentice, A. M., and Paul, A. A. (1994) Seasonal variations in vitamins A, C, riboflavin and folate intakes and status of pregnant and lactating women in a rural Gambian community: some possible implications. *Eur. J. Clin. Nutr.* **48,** 660–668.
18. Becker, K. and Schirmer, R. H. (1990) The EGRAC as a measure of the riboflavin status in man. Titration of hemolysate FAD with apoglutathione reductase. *Flavins Flavoproteins* **10,** 851–854.
19. Kapur, S., Ganguli, P., Ulrich, R., and Raghu, U. (1991) Use of random-sequence riboflavin as a marker of medication compliance in chronic schizophrenics. *Schizophr. Res.* **6,** 49–53.
20. Hungerbuhler, P., Bovet, P., Shamlaye, C., Burnand, B., and Waeber, B. (1995) Compliance with medication among outpatients with uncontrolled hypertension in the Seychelles. *Bull. World Health Organ.* **73,** 437–442.

21. Del-Boca, F. K., Kranzler, H. R., Brown, J., and Korner, P. F. (1996) Assessment of medication compliance in alcoholics through UV light detection of a riboflavin tracer. *Alcohol Clin. Exp. Res.* **20,** 1412–1417.
22. Davies, M. G., Fulton, G. J., and Hagen, P. O. (1995) Clinical biology of nitric oxide. *Br. J. Surg.* **82,** 1598–1610.
23. Kröncke, K. D., Fehsel, K., and Kolb-Bachofen, V. (1997) Nitric oxide. Cytotoxicity versus cytoprotection—how, why, when and where? *NITRIC OXIDE Biol. Chem.* **1,** 107–120.
24. Becker, K., Gui, M., and Schirmer, R. H. (1995) Inhibition of human glutathione reductase by *S*-nitrosoglutathione. *Eur. J. Biochem.* **234,** 472–478.
25. Keese, M., Boese, M., Mülsch, A., Schirmer, R. H., and Becker, K. (1997) Dinitrosyl-dithiol-iron complexes, NO-carriers *in vivo*, act as potent inhibitors of human glutathione reductase and glutathione S-transferase. *Biochem. Pharmacol.* **54,** 1307–1313.
26. Boese, M., Keese, M., Becker, K., Busse, R., and Mülsch, A. (1997) Inhibition of glutathione reductase by dinitrosyl-iron-dithiolate complex. *J. Biol. Chem.* **272,** 21,767–21,773.
27. Karplus, P. A. and Schulz, G. E. (1987) Refined structure of glutathione reductase at 1.54 Å resolution. *J. Mol. Biol.* **195,** 701–729.
28. Färber, P. M., Becker, K., Müller, S., Schirmer, R. H., and Franklin, R. M. (1996) Molecular cloning and characterization of a putative glutathione reductase gene, the PfGR2 gene, from *Plasmodium falciparum. Eur. J. Biochem.* **239,** 655–661.
29. Becker K., Färber P. M., von der Lieth C. W., and Müller S. (1997) Glutathione reductase and thioredoxin reductase of the malaria parasite *Plasmodium falciparum. Flavins Flavoproteins* **12,** 13–22.
30. Schirmer, M. (1998) Dimerisierungsinhibitoren der flavinhaltigen Disulfidreduktasen aus malariainfizierten Erythrozyten. MD-thesis. Heidelberg University, Heidelberg, Germany.
31. Nordhoff, A., Tziatzios, C., Van den Broek, A., Schott, M. K., Kalbitzer, H.-R., Becker, K., Schubert, D., and Schirmer, R. H. (1997) Denaturation and reactivation of dimeric human glutathione reductase. An assay for folding inhibitors. *Eur. J. Biochem.* **245,** 273–282.
32. Pongpaew, P., Saowakontha, S., Schelp, F. P., Rojsathaporn, K., and Phonrat, B. (1995) Vitamin B_1, B_2 and B_6 during the course of pregnancy of rural and urban women in northeast Thailand. *Int. J. Vitam. Nutr. Res.* **65,** 111–116.
33. Flatz, G. (1971) Population study of erythrocyte glutathione reductase activity. *Humangenetik* **11,** 269–277.
34. Becker, K., Leichsenring, M., Gana, L., Bremer, H. J., and Schirmer, R. H. (1995) Glutathione and associated antioxidant systems in protein energy malnutrition: results of a study in Nigeria. *Free Rad. Biol. Med.* **18,** 257–263.
35. Tucker, R. G., Mickelsen, O., and Keys, A. (1960) The influence of sleep, work, diuresis, heat, acute starvation, thiamine intake, and bed rest on human riboflavin starvation. *J. Nutr.* **72,** 251–261.

36. Lee, S. S. and McCormick, D. B. (1985) Thyroid hormone regulation of flavocoenzyme biosynthesis. *Arch. Biochem. Biophys.* **237,** 197–201.
37. Pinto, J., Huang, Y. P., and Rivlin, R. S. (1985) Inhibition by chlorpromazine of thyroxine modulation of flavin metabolism in liver, cerebrum and cerebellum. *Biochem. Pharmacol.* **34,** 93–95.
38. Becker, K. and Wilkinson, A. (1993) Flavin adenine dinucleotide levels in erythrocytes of very low birthweight infants under vitamin supplementation. *Biol. Neonate* **63,** 80–85.
39. Baeckert, P. A., Greene, H. L., Fritz, I., Oelberg, D. G., and Adcock, E. W. (1988) Vitamin concentrations in very low birth weight infants given vitamins intravenously in a lipid emulsion: measurement of vitamins A, D, and E and riboflavin. *J. Pediatr.* **113,** 1057–1065.
40. Joshi, P. C. (1989) Ultraviolet radiation-induced photodegradation and $^1O_2$, O_2^- production by riboflavin, lumichrome and lumiflavin. *Indian J. Biochem. Biophys.* **26,** 186–189.
41. Becker, K. and Schirmer, R. H. (1995) 1,3-Bis(2-chloroethyl)-1-nitrosourea as thiol-carbamoylating agent in biological systems. *Methods Enzymol.* **251,** 173–188.
42. Ogle, R. F., Christodoulou, J., Fagan, E., Blok, R. B., Kirby, D. M., Seller, K. L., Dahl, H. H., and Thorburn, D. R. (1997) Mitochondrial myopathy with tRNA(Leu(UUR)) mutation and complex I deficiency responsive to riboflavin. *J. Pediatr.* **130,** 138–145.
43. Mack, C. P., Hultquist, D. E., and Shlafer, M. (1995) Myocardial flavin reductase and riboflavin: a potential role in decreasing reoxygenation injury. *Biochem. Biophys. Res. Comm.* **212,** 35–40.
44. Natraj, U., George, S., and Kadam, P. (1988) Isolation and partial characterization of human riboflavin carrier protein and the estimation of its level during human pregnancy. *J. Reprod. Immunol.* **13,** 1–7.
45. Adiga, P. R., Subramanian, S., Rao, J., and Kumar, M. (1997) Prospects of riboflavin carrier protein (RCP) as an antifertility vaccine in male and female mammals. *Hum. Reprod. Update* **3,** 299–300.
46. Scholte, H. R., Busch, H. F., Bakker, H. D., Bogaard, J. M., Luyt-Houwen, I. E., and Kuyt, L. P. (1995) Riboflavin-responsive complex I deficiency. *Biochim. Biophys. Acta* 1**271,** 75–83.
47. Kmoch, S., Zeman, J., Hrebicek, M., Ryba, L., Kristensen, M. J., and Grgersen, N. (1995) Riboflavin-responsive epilepsy in a patient with SER209 variant form of short-chain acyl-CoA dehydrogenase. *J. Inherit. Metab. Dis.* **18,** 227–229.
48. Schoenen, J. (1997) Migraine: genetic, physiopathological, and therapeutic innovations. *Rev. Med. Liege.* **52,** 83–88.
49. Prasad, M. P., Mukudan, M. A., and Krishnaswamy, K. (1995) Micronuclei and carcinogen DNA adducts as intermediate end points in nutrient intervention trial of precancerous lesions in the oral cavity. *Eur. J. Cancer. Oral. Oncol.* **31B,** 155–159.
50. Webster, R. P., Gawde, M. D., and Bhattacharya, R. K. (1996) Modulation of carcinogen-induced DNA damage and repair enzyme activity by dietary riboflavin. *Cancer Lett.* **98,** 129–135.

51. Khadem, J., Truong, T., and Ernest, J. T. (1994) Photodynamic biologic tissue glue. *Cornea* **13,** 406–410.
52. Keese, M. (1997) Zur Zellbiologie der Glutathionreduktase: Wirkung physiologischer NO-Carrier und Entwicklung eines Indikatorsystems für den zellulären Redoxmetabolismus. MD thesis. Heidelberg University.
53. Schirmer, R. H., Krauth-Siegel, R. L., and Schulz, G. E. (1989) Glutathione reductase, in *Coenzymes and Cofactors*, vol. 3A (Dolphin, D., Poulson, R., and Avramovic, O., eds.), Wiley, New York, pp. 545–596.
54. Arscott, L. D., Gromer, S., Schirmer, R. H., Becker, K., and Williams, C. H. (1997) The mechanism of thioredoxin reductase from human placenta is similar to the mechanisms of lipoamide dehydrogenase and glutathione reductase and is distinct from the mechanism of thioredoxin reductase from *Escherichia coli. Proc. Natl. Acad. Sci. USA.* **94,** 3621–3626.
55. Murthy, Y. V. and Massey, V. (1995) Chemical modification of the *N*-10 ribityl side chain of flavins. Effects on properties of flavoprotein disulfide oxidoreductases. *J. Biol. Chem.* **270,** 28,586–28,594.
56. Krauth-Siegel, R. L., Schirmer, R. H., and Ghisla, S. (1985) FAD analogues as prosthetic groups of human glutathione reductase. Properties of the modified enzyme species and comparisons with the active site structure. *Eur. J. Biochem.* **148,** 335–344.
57. Cowden, W. B., Clark, I. A., and Hunt N. H. (1988) Flavins as potential antimalarials. 1. 10-(Halophenyl)-3-methylflavins. *J. Med. Chem.* **31,** 799–801.
58. Becker, K., Christopherson, R. J., Cowden, W. B., Hunt, H. N., and Schirmer, R. H. (1990) Flavin analogs with antimalarial activity as glutathione reductase inhibitors. *Biochem. Pharmacol.* **39,** 59–65.
59. Schönleben-Janas, A., Kirsch, P., Mittl, P. R. E., Schirmer, R. H., and Krauth-Siegel, R. L. (1996) Inhibition of human glutathione reductase by 10-arylisoalloxazines: crystalline, kinetic, and electrochemical studies. *J. Med. Chem.* **39,** 1549–1555.
60. Williams, C. H., Jr (1992) Lipoamide dehydrogenase, glutathione reductase, thioredoxin reductase, and mercuric ion reductase—a family of flavoenzyme transhydrogenases, in *Chemistry and Biochemistry of Flavoenzymes*, vol. 3 (Müller, F., ed.), CRC Press, Boca Raton, FL, pp. 121–211.

Index

G

H

I

L

M

N

O